Bridge Over Troubled Waters

LINKING CLIMATE CHANGE AND DEVELOPMENT

ORGANISATION FOR ECONOMIC CO-OPERATION AND DEVELOPMENT

ORGANISATION FOR ECONOMIC CO-OPERATION AND DEVELOPMENT

The OECD is a unique forum where the governments of 30 democracies work together to address the economic, social and environmental challenges of globalisation. The OECD is also at the forefront of efforts to understand and to help governments respond to new developments and concerns, such as corporate governance, the information economy and the challenges of an ageing population. The Organisation provides a setting where governments can compare policy experiences, seek answers to common problems, identify good practice and work to co-ordinate domestic and international policies.

The OECD member countries are: Australia, Austria, Belgium, Canada, the Czech Republic, Denmark, Finland, France, Germany, Greece, Hungary, Iceland, Ireland, Italy, Japan, Korea, Luxembourg, Mexico, the Netherlands, New Zealand, Norway, Poland, Portugal, the Slovak Republic, Spain, Sweden, Switzerland, Turkey, the United Kingdom and the United States. The Commission of the European Communities takes part in the work of the OECD.

OECD Publishing disseminates widely the results of the Organisation's statistics gathering and research on economic, social and environmental issues, as well as the conventions, guidelines and standards agreed by its members.

Publié en français sous le titre :
Contre vents et marées
LES POLITIQUES DE DÉVELOPPEMENT FACE AU CHANGEMENT CLIMATIQUE

ISBN 92-64-01275-3
Bridge Over Troubled Waters
Linking Climate Change and Development

Editor

Shardul Agrawala (OECD)

Contributors

Ahsan Uddin Ahmed (Bangladesh Unnayan Parishad, Bangladesh)
Walter Baethgen (Columbia University, USA)
Declan Conway (University of East Anglia, UK)
Mohamed El Raey (University of Alexandria, Egypt)
Frédéric Gagnon-Lebrun (ÉcoRessources Consultants, Canada)
Simone Gigli (OECD)
Andreas Hemp (University of Bayreuth, Germany)
Peter Larsen (National Center for Atmospheric Research, USA)
Daniel Martino (Carbosur, Uruguay)
Annett Moehner (UNFCCC)
Vivian Raksakulthai (Asian Disaster Preparedness Center, Thailand)
James Risbey (CSIRO Marine and Atmospheric Research, Australia)
Joel Smith (Stratus Consulting, USA)
Maarten van Aalst (Utrecht University, The Netherlands)

Foreword

Climate change poses a serious challenge to social and economic development in all countries. Developing countries are particularly vulnerable because of their high dependence on natural resources and their limited capacity to cope with the impacts of climate change. Clearly, while we continue the process of negotiating international commitments to reduce greenhouse gas emissions, that alone will not be enough. We also need to place climate change and its impacts into the mainstream of our economic policies, development projects and international aid efforts.

It is within this context that the OECD Environment and Development Co-operation directorates undertook a major collaborative project on mainstreaming responses to climate change in development planning and assistance. A primary focus of the work has been on adaptation to climate change, although links between greenhouse gas mitigation and development objectives were also considered. Six case studies were conducted in Bangladesh, Egypt, Fiji, Nepal, Tanzania and Uruguay as part of this effort by an international team of experts. These studies have now all been completed and are available at www.oecd.org/env/cc.

This volume highlights cross-cutting findings and key methodological issues from the country case studies. Together, the authors of this volume underscore the need for mainstreaming of climate change considerations in development activities as well as the challenges involved. Overall the volume suggests a rich agenda for research and policy action which should be of considerable interest to donor agencies, sectoral planners and development experts, as well as to climate change experts and policy makers.

Kiyo Akasaka
Deputy Secretary-General
OECD

Acknowledgments. *The OECD would like to thank member governments as well as many external partners that have made this book possible. The project was overseen by the Working Party on Global and Structural Policies (WPGSP) of the Environmental Policy Committee and the Network on Environment and Development Co-operation (Environet) of the Development Assistance Committee. This volume has greatly benefited from comments from delegates to the WPGSP and Environet.*

In addition to the authors of this volume, valuable contributions to the underlying country case studies were made by Kanyathu Koshy (University of South Pacific, Fiji), Eugenio Lorenzo (Universidad de la Republica, Uruguay), John Reynolds (Reynolds Geosciences, UK), Arun Shreshtha (Department of Hydrology and Meteorology, Nepal) and the Center for Energy, Environment, Science and Technology (Dar es Salaam, Tanzania). This work has also benefited from inputs from experts at a workshop organised by the Government of Nepal in March 2003 and a scoping workshop at OECD headquarters in March 2002.

The project could not have been completed without the financial support received from the governments of Canada, Germany, Ireland, Japan, Korea, Norway and the Netherlands. This support is gratefully acknowledged.

Shardul Agrawala of the OECD Environment Directorate edited this book and oversaw the project leading up to it. Valuable feedback and guidance was provided by Remy Paris and Georg Caspary of the OECD Development Co-operation Directorate, and Tom Jones and Jan Corfee-Morlot of the OECD Environment Directorate. Simone Gigli, Anne de Montfalcon, Annett Moehner, Frédéric Gagnon-Lebrun, Carolyn Sturgeon-Bodineau, Tomoko Ota and Martin Berg provided invaluable staff support for the book and the project. Rebecca Brite provided painstaking editorial support and Jonas Franke produced GIS maps under a very tight deadline.

Photo Credits

Nepal

Plates 1 and 2	Photos courtesy of Shreekamal Dwivedi, Department of Water Induced Disaster Prevention, Nepal.
Plate 3	Photo by Saskia Clark.
Plates 4 and 5	Photos courtesy of Department of Hydrology and Meteorology, Nepal.

Tanzania

Plates 6, 7, 8 and 9	Photos by Andreas Hemp, University of Bayreuth.
Plates 10 and 11	Maps courtesy of United Nations Environment Program and the University of Bayreuth.

Egypt

Plate 13	Global Land Cover 2000 database. European Commission, Joint Research Centre, 2003, *www.gvm.jrc.it/glc2000*.
Plate 14	Center for International Earth Science Information Network (CIESIN), Columbia University; and Centro Internacional de Agricultura Tropical (CIAT) 2004. Gridded Population of the World (GPW), Version 3. Palisades, NY: CIESIN, Columbia University. Available at *http://sedac.ciesin.columbia.edu/gpw*.

Fiji

Plate 15	Photo by James Risbey.
Plate 16	Photo courtesy of WWF, South Pacific Programme.

Uruguay

Plates 17 and 18	Photos by Daniel Martino, Carbosur, Uruguay.

Table of Contents

List of boxes

BRIDGE OVER TROUBLED WATERS – ISBN 92-64-01275-3 – © OECD 2005

List of tables

List of figures

List of Abbreviations

ADB	Asian Development Bank
CBD	Convention on Biological Diversity
CDM	Clean Development Mechanism
COP	Conference of the Parties
CRS	Creditor Reporting System
DAC	Development Assistance Committee (of the OECD)
EIA	Environmental impact assessment
EU	European Union
GCM	General circulation model
GEF	Global Environment Facility
GHG	Greenhouse gas
GLOF	Glacial lake outburst flood
GTZ	Gesellschaft für Technische Zusammenarbeit (German development agency)
IPCC	Intergovernmental Panel on Climate Change
LDCF	Least Developed Countries Fund
LDCs	Least developed countries
MDG	Millennium Development Goal
NAPA	National Adaptation Programme of Action
NGO	Non-governmental organisation
NORAD	Norwegian Agency for Development Co-operation
ODA	Official Development Assistance
OOF	Other official flows
PRSP	Poverty Reduction Strategy Paper
SCCF	Special Climate Change Fund
SRES	Special Reports on Emission Scenarios
UMICs	Upper middle income countries
UNCCD	United Nations Convention to Combat Desertification
UNFCCC	United Nations Framework Convention on Climate Change

ISBN 92-64-01275-3
Bridge Over Troubled Waters
Linking Climate Change and Development

Executive Summary

The issue of climate change can seem remote, compared with such immediate problems as poverty, disease and economic stagnation. Yet, climate change can directly affect the efficiency of resource investments and eventual achievement of many development objectives. How development occurs also has implications for climate change itself and the vulnerability of societies to its impacts. There is therefore a need to link climate change considerations with development priorities.

Considerable analytical work has already been done on how development can be made climate-friendly in terms of helping reduce greenhouse gas emissions which cause climate change, although implementation remains a challenge. Much less attention has been paid to how development can be made more resilient to the impacts of climate change. In a narrow engineering sense, it could, for example, involve taking into account impacts such as sea level rise and glacial lake outburst floods in the siting and design of bridges and other infrastructure. At a policy level it could involve considering the implications of climate change on a variety of development activities including poverty reduction, sectoral development, and natural resource management. Bridging the gap between the climate change and development communities, however, requires more than a simple dialogue. This is because they have different priorities, often operate on different time and space scales, and do not necessarily speak the same language. Specific information is therefore needed on the significance of climate change for development activities along with operational guidance on how best to respond to it within the context of other pressing social priorities.

This volume synthesises the results of an OECD project on the opportunities and trade-offs faced in "mainstreaming" responses to climate change in development planning and assistance. Six country case studies reviewed climate change impacts and vulnerabilities, analysed relevant national plans and aid portfolios and examined in depth selected areas of natural resource management where climate change is closely intertwined with development: water resource management on the Nile in Egypt, coastal mangroves in Fiji and Bangladesh, glacier retreat and water resource management in Nepal, economic development and natural resource management on Mount Kilimanjaro in Tanzania, and forestry and agriculture in Uruguay. A primary focus of this work was on mainstreaming adaptation to the impacts of climate change, although links between greenhouse gas mitigation, natural resource management and development priorities were also considered.

A summary assessment

Several findings have emerged from this work which reinforce the need for, and the challenges faced in, taking climate change into account in development planning and activities.

Climate change is already affecting development

In addition to natural climate variability, long-term trends and climate change are already having a discernible impact on development. This is particularly the case for the impacts of glacier retreat and increased risk of glacial lake outburst flooding which are closely related to observed trends in rising temperatures. Clearly, a diverse range of development activities, from design of hydropower facilities to rural development and settlement policies, will need to adapt to the impacts of both current and future climate risks.

Future climate change impacts may also need consideration in development planning

Even in cases where the impacts of climate change are not yet discernible, scenarios of future impacts may already be sufficient to justify building some adaptation responses into planning. One reason is that it could be more cost-effective to implement adaptation measures early on, particularly for long-lived infrastructure. Another reason is that, in many contexts, current development activities may irreversibly constrain future adaptation to the impacts of climate change. This could be true, for example, in the case of destruction of coastal mangroves, or development of human settlements in areas that are likely to be particularly exposed to climate change. In such instances, even near term policies may need to consider the long-term implications of climate change.

A significant portion of development assistance is directed at climate-sensitive activities

An analysis of the composition of Official Development Assistance flows to the six case study countries indicates that a significant portion is directed at activities potentially affected by climate risks, including climate change. Expressed as a percentage of total national official flows, estimates range from as high as 50-65 per cent in Nepal to 12-26 per cent in Tanzania. While any classification of this nature suffers from oversimplification, the analysis underscores the fact that consideration of climate risks is often important for development investments and projects.

Development activities routinely overlook climate change and often even climate variability

Some weather and climate considerations are routinely taken into account in a wide range of activities, from crop selection to the design of highways and energy generation facilities. However, not all climate risks are being incorporated in decision making, even with regard to natural weather extremes. Nor are practices that take into account historical climate necessarily suitable under climate change. Many planning decisions focus on shorter timescales and tend to neglect the longer-term perspective. An analysis of national development plans, poverty reduction strategy papers, sectoral strategies and project documents in climate-sensitive sectors indicates that such documents generally pay little or no attention to climate change, and often only limited attention to current climate risk. Even when climate change is mentioned, specific operational guidance on how to take it into account is generally lacking.

Barriers to mainstreaming climate change

Why is it so difficult to implement and mainstream responses to climate change – particularly adaptation – within development activity? Lack of awareness of climate change within the development community and limitations on resources to implement response measures are the most frequently cited explanations. They may well hold true in many situations, but underlying them is a more complex web of reasons.

Segmentation and other barriers within governments and donor agencies limit mainstreaming

Climate change expertise is typically housed in environment departments of governments and donor agencies which have limited leverage over sectoral guidelines and projects. Sectoral managers and country representatives may also face "mainstreaming overload", with competing agendas such as gender, governance and environment vying for integration within core development activities. Many development projects continue to be funded over three to five year time horizons, and as such may not be the best vehicle for long-term climate risk reduction. Adaptation to climate change may also have more difficulty attracting resources than more visible activities such as emergency response, post-disaster recovery and reconstruction, where funding modalities are better established.

Available climate information is often not directly relevant for development-related decisions

Development activities are sensitive to a broad range of climate variables – only some of which can be reliably projected by climate models. Temperature, for example, is typically easier to project than rainfall. Climate extremes, which are often critical for many development-related decisions, are much more difficult to project than mean trends. There is also a mismatch between the time and space scales of climate change projections and the information needs of development planners. For example, the primary sensitivity of development activities to climate is at a local scale (such as a watershed or a city), for which credible climate change projections are often lacking.

Sometimes there are trade-offs between climate and development objectives

Mainstreaming could also prove difficult to carry out because of direct trade-offs in certain cases between development priorities and the actions required to deal with climate change. Governments and donors confronting pressing challenges, such as poverty and inadequate infrastructure, have few incentives to divert scarce resources to investments that are perceived as not paying off until climate change impacts fully manifest themselves. Putting a real value on natural resources and deciding when *not* to develop coastal areas or hillsides may also be seen as hampering development. At the project level, mainstreaming can be thought of as complicating operating procedures with additional requirements or considerations, or raising costs. In addition, short-term economic benefits that often accrue to a few in the community can crowd out longer-term considerations such as climate change. Shrimp farming, mangrove conversion and infrastructure development, for example, provide employment and boost incomes, but they may also promote maladaptation and increase the vulnerability of critical coastal systems to climate change impacts.

BRIDGE OVER TROUBLED WATERS – ISBN 92-64-01275-3 – © OECD 2005

Opportunities for the road ahead

Several opportunities exist for more effective integration of climate change considerations within development activities.

Making climate information more relevant and usable

Development practitioners need access to credible and context-specific climate information as a basis for decisions. This includes information on the cost and effectiveness of integrating adaptation or mitigation measures within development planning. Perhaps even more fundamental in the case of adaptation is information on the impacts of climate change and variability on particular development activities. While it would be naïve to call for a significant reduction in scientific uncertainty in climate model projections, more can be done to facilitate transparent communication of this uncertainty to development practitioners. Analysis of the costs and distributional aspects of adaptation could also assist sectoral decision makers in determining the degree to which they should integrate such responses within their core activities.

Developing and applying climate risk screening tools

In addition to improving the quality of climate information, tools and approaches are needed to assess the potential exposure of a broad range of development activities to climate risks and to prioritise responses. Also needed are more sophisticated screening tools at the project level, in order to identify the key variables of relevance to the project, how they are affected by climate change and what implications this has on the viability of the project. Field-testing of such screening tools and their diffusion to a wide range of project settings could greatly advance the integration of climate risks in development activities.

Identifying and using appropriate entry points for climate information

Identification of appropriate entry points for climate change information in development activities is greatly needed. Potential entry points for the use of climate information and for integrating adaptation include land use planning, disaster response strategies and infrastructure design. Environmental Impact Assessments (EIAs) could be another entry point for mainstreaming both mitigation and adaptation. The implications of projects for greenhouse gas emissions could be included in EIA checklists. However, EIA guidelines would need to be broadened to include consideration of climate change impacts. This is because current guidelines only consider the impact of a project or activity *on* the environment, and not the impact of the environment on the project. It is also important to embed climate change considerations within planning mechanisms and ensure that the responsibility for co-ordination lies with an influential department. Furthermore, attention should be given not only to investment plans but also to legislation.

Shifting emphasis to implementation, as opposed to developing new plans

In many instances, rather than requiring radically new responses, climate change only reinforces the need for implementation of measures that already are, or should be, environmental or development priorities. Examples include water or energy conservation, forest protection and afforestation, flood control, building of coastal embankments, dredging to improve river flow and protection of mangroves. Often such measures have already been called for in national and sectoral planning documents but not successfully implemented. Reiteration of the measures in elaborate climate change plans is unlikely to have much effect on the ground unless barriers to effective implementation of the existing sectoral and development plans are confronted. Putting the spotlight on implementation, therefore, could put the focus on greater accountability in action on the ground.

Encouraging meaningful co-ordination and the sharing of good practices

Institutional mechanisms need to be developed to forge links between mainstreaming initiated under the international climate change regime and the risk management activities of national and sectoral planners. A corollary link could be between activities initiated to achieve development objectives, such as the Millennium Development Goals, and more bottom-up consideration of the impacts of climate change. Greater engagement of the private sector and local communities in mainstreaming efforts is also needed. Another priority that has not received sufficient attention is trans-boundary and regional co-ordination. Most climate change action and adaptation plans are at the national level while many impacts cut across national boundaries. Meaningful integration of a range of climate risks, from flood control to dry season flows to glacial lake hazards, would require greater co-ordination on data collection, monitoring and policies at the regional level. Finally, operational guidance on comprehensive climate risk management in development is needed to facilitate policy coherence, allow for joint building of experience and promote sharing of tools and experiences within and among governments and development co-operation agencies.

ISBN 92-64-01275-3
Bridge Over Troubled Waters
Linking Climate Change and Development

Chapter 1

Putting Climate Change in the Development Mainstream: Introduction and Framework

by

Shardul Agrawala

Climate change can appear remote compared with problems such as poverty, disease and economic stagnation. Development planners are often unsure how it will affect their work, and whether and how to integrate or "mainstream" climate change considerations within their activities. Climate change is in fact intricately tied to many development objectives. Furthermore, how development occurs has implications for climate change as well as the vulnerability of societies to its impacts. This chapter outlines the links between climate change and development. It defines key concepts including weather, climate variability and climate change, and response measures, particularly adaptation to climate change impacts and mitigation of greenhouse gas emissions. The chapter then outlines the framework of case studies of Bangladesh, Egypt, Fiji, Nepal, Tanzania and Uruguay, which were conducted as part of an OECD project on mainstreaming responses to climate change in development planning and assistance.

1. Introduction

Governments and policy makers worldwide continue to search for new solutions to pressing development challenges such as persistent poverty, malnutrition, spread of infectious diseases, natural hazards, unemployment and economic stagnation. The social impacts of such problems are typically all too evident, as is the need for policy responses. Against this backdrop of the visible and urgent, climate change often looms as a somewhat esoteric, long-term, global threat to the development community. Development planners are often uncertain about its implications on their particular region or locus of activities, and whether and how to incorporate or "mainstream" considerations of climate change within their planning and project activities.

This volume synthesises the results of an OECD project on mainstreaming responses to climate change in development planning and assistance. Jointly initiated by the OECD Environment and Development Co-operation directorates, the project had as its main objective to explore possible synergies and trade-offs in mainstreaming responses to climate change in development assistance policies of donor agencies, as well as in national and sectoral policies and projects of recipient countries. Six case studies were conducted as part of this project, in Bangladesh, Egypt, Fiji, Nepal, Tanzania and Uruguay, and published separately (OECD, 2003a-d; OECD, 2004a-b). Each case study followed a three-tier framework: a country-level overview of the principal effects of climate change and vulnerabilities to it; analyses of national plans and development assistance portfolios that bear upon vulnerable sectors and regions; and in-depth regional or sectoral analyses of how climate change adaptation responses can be mainstreamed in particular development policies and projects (Agrawala and Berg, 2002). The in-depth analyses addressed a range of issues in which climate change adaptation is very closely intertwined with development, such as water resource management on the Nile, coastal mangroves in Fiji and the Bangladesh Sundarbans, glacier retreat and water resource management in Nepal, economic development and natural resource management on Mount Kilimanjaro and coastal zone management in Egypt and Uruguay.

This introductory chapter provides an overview of key issues related to links between climate change and development, and then discusses the analytical framework for the case studies. The following section introduces human-induced climate change and places it within the context of weather and current climate variability. Next, key links between climate change and development are reviewed, followed by a discussion of mitigation of and adaptation to climate change and the challenge of mainstreaming such measures within business-as-usual development activities. The chapter then focuses on adaptation, reviewing the status of related policy initiatives within the international climate change regime as well as early efforts to mainstream adaptation in development activities. The remaining sections outline the objectives and scope of this project and provide an overview of the analytical framework and case study countries.

2. Weather, climate variability and climate change

Weather describes the actual state of the atmosphere at a given location in terms of such variables as air temperature, pressure, humidity and wind speed. *Climate* is typically defined as weather averaged over a period of time and possibly over a geographic region. Climate is what one expects, weather is what one gets.

A look at various climate records shows variability on every timescale, from daily, seasonal and annual to over hundreds or even thousands of years. Until very recently the causes of such changes were entirely natural. They included the chaotic dynamics inherent to the climate system as well as changes in solar radiation, ocean circulation and reflectance of the earth's surface. Phenomena as diverse as the last ice age (which ended 10 000 years ago) and shorter-term fluctuations driven by the El Niño Southern Oscillation (which occur every couple of years and typically last several months) are examples of natural shifts in the climate system.

Upon this complex landscape of naturally occurring climatic changes and fluctuations, human activity has superimposed a relatively recent trend: anthropogenic climate change or "global warming". There is now an international consensus among the world's leading experts, assembled under the auspices of the Intergovernmental Panel on Climate Change (IPCC), that increases in concentrations of greenhouse gases (GHGs) since the dawn of the industrial revolution have already had an influence on climate that became discernible during the 20th century (IPCC, 1995a and 2001a). The mean global temperature increased by almost 0.6 °C over the course of that century, with most of the warming coming in the last few decades following a sharp increase in GHG concentrations since the 1950s. Citing scenarios of future GHG emissions and projections from numerical climate models, the IPCC estimates that the earth's surface temperature, globally averaged, will increase by between 1.4 °C and 5.8 °C over 1990 to 2100. It is likely that such a rate of warming has not occurred in at least a thousand years. The IPCC also projects that global mean sea levels will rise by at least 9 cm and perhaps as much as 88 cm as a result of thermal expansion of oceans and the melting of ice caps.

Climate change is linked not only to mean changes in variables such as temperature, but also to changes in certain extremes, which often have much more significant consequences on society. The IPCC Third Assessment Report provides an assessment of confidence in observed changes in extremes of weather and climate during the 20th century and in projected changes during the 21st century (Table 1.1). Temperature-related extremes – higher likelihood of heat waves and reduced likelihood of cold waves – have the closest and most generalisable link to climate change. Climate change might also result in more intense precipitation, though that has thus far been observed only in mid- to high-latitude areas in the northern hemisphere. Similarly, increased continental drying and higher risk of drought are projected, but only over mid-latitude continental interiors. Increases in tropical cyclone intensity are projected in some areas, but have not been demonstrably observed during the 20th century. Potential links between other weather and climate extremes and climate change have not yet been well established (IPCC, 2001a).

Table 1.1. **Estimates of confidence in observed and projected changes in extreme weather and climate events**

Confidence in observed changes (latter half of the 20th century)	Changes in phenomenon	Confidence in projected changes (21st century)
Likely	*Higher maximum temperatures and more hot days over nearly all land areas*	Very likely
Very likely	*Higher minimum temperatures and fewer cold days and frost days over nearly all land areas*	Very likely
Very likely	*Reduced diurnal temperature range over most land areas*	Very likely
Likely, over many areas	*Increase of heat index over land areas*	Very likely, over most areas
Likely, over many northern hemisphere mid- to high-latitude land areas	*More intense precipitation events*	Very likely, over most areas
Likely, in a few areas	*Increase in summer continental drying and associated risk of drought*	Likely, over most mid-latitude continental interiors. (Lack of consistent projections in other areas)
Not observed in the few analyses available	*Increase in tropical cyclone peak wind intensities*	Likely, over some areas
Insufficient data for assessment	*Increase in tropical cyclone mean and peak precipitation intensities*	Likely, over some areas

Source: IPCC (2001a).

3. Climate change and development: key links

Unquestionably, climate is closely intertwined with development choices and pathways. Climate is a resource in itself, and it mediates the productivity of

BRIDGE OVER TROUBLED WATERS – ISBN 92-64-01275-3 – © OECD 2005

other critical resources, including food and fibre, forests, fisheries and water resources. Climate can also be a hazard. Natural fluctuations such as those related to El Niño cause widespread disruptions in society's ability to harness critical resources and even to survive. It is equally the case that human development choices are having a demonstrable impact on local and global climate patterns. Over-construction contributes to the formation of local urban "heat islands"; deforestation and land use changes can influence regional temperature and rainfall patterns; and, as already noted, increases in GHG concentrations as a result of industrial activity are responsible for global climate change. Developing countries contribute 34% of global CO_2 emissions today, but if current trends in economic development continue, by 2030 they will account for 47%, nearly half of global emissions (IEA, 2002). Development choices and pathways in developing countries would thus be a critical vehicle for future GHG emission reductions to complement reductions in the industrialised world.

Climate change in turn will have attendant impacts on a variety of natural and human systems worldwide in the decades and centuries to come, affecting the development potential of future generations. Meanwhile, the vulnerability of particular systems to the effects of climate change depends not only on the magnitude of climatic stress, but also on their sensitivity and capacity to adapt to or cope with such stress (Box 1.1).

Box 1.1. **Climate change sensitivity, adaptive capacity and vulnerability**

Sensitivity is the degree to which a system is affected, negatively or positively, by climate-related stimuli. Such stimuli encompass all elements of climate change, including mean climate characteristics, climate variability and the frequency and magnitude of extremes. The effect may be direct (for example, a change in crop yield in response to a change in the mean, range or variability of temperature) or indirect (such as damage caused by increased frequency of coastal flooding due to sea-level rise).

Adaptive capacity is a system's ability to adjust to climate change (including climate variability and extremes), to moderate potential damage, to take advantage of opportunities or to cope with consequences.

Vulnerability is the degree to which a system is susceptible to, or unable to cope with, adverse effects of climate change, including climate variability and extremes. Vulnerability is a function of the character, magnitude and rate of climate change, and the variation to which a system is exposed, along with its sensitivity and adaptive capacity.

Source: IPCC (2001b).

Sensitivity to climatic stress is higher for activities entailing climate-dependent natural resources, such as agriculture and coastal resources – often critical for the livelihoods of the poor. The capacity to adapt and cope depends upon such factors as wealth, technology, education, information, skills and access to resources – all scarce in poor countries and communities. The higher sensitivity of livelihoods to climate effects, combined with lower capacity to cope, renders developing countries disproportionately vulnerable to climate change (IPCC, 2001b).

Consequently, the effects of climate change may also be critical to the achievement of many development objectives, particularly as they relate to the most vulnerable groups and communities. The projected impact on access to natural resources, heat-related mortality and spread of vector-borne diseases, for example, has direct implications for several of the Millennium Development Goals (MDGs), which all member countries of the United Nations have adopted as a road map for development efforts (Sperling, 2003: Table 1.2). The target date for achieving the MDGs, however, is 2015, and the effects of climate change will continue to emerge – in fact, will become progressively more significant – in the years and decades beyond 2015.

Table 1.2. **Potential implications of climate change for Millennium Development Goals**

Millennium Development Goal	Examples of links with climate change
Eradicate extreme poverty and hunger (Goal 1)	Climate change is projected to reduce poor people's livelihood assets; for example, health, access to water, homes, and infrastructure. Climate change is expected to alter the path and rate of economic growth due to changes in natural systems and resources, infrastructure and labor productivity. A reduction in economic growth directly impacts poverty through reduced income opportunities. Climate change is projected to alter regional food security. In particular in Africa, food security is expected to worsen.
Health related goals: Combat major diseases Reduce infant mortality Improve maternal health (Goals 4, 5 and 6)	Direct effects of climate change include increases in heat-related mortality and illness associated with heatwaves (which may be balanced by reductions in winter cold-related deaths in some regions). Climate change may increase the prevalence of some vector-borne diseases (such as malaria and dengue fever) and vulnerability to ailments transmitted via water, food or human proximity (*e.g.* cholera, dysentery). Children and pregnant women are particularly susceptible to vector- and waterborne diseases. Anaemia resulting from malaria is responsible for one-fourth of maternal mortality. Climate change is likely to result in declining quantity and quality of drinking water, which is a prerequisite for good health, and to exacerbate malnutrition – an important source of ill health among children – by reducing natural resource productivity and threatening food security, particularly in sub-Saharan Africa.
Ensure environmental sustainability (Goal 7)	Climate change is projected to alter the quality and productivity of natural resources and ecosystems, some of which may be irreversibly damaged, and these changes may also decrease biological diversity and compound existing environmental degradation.

Source: Sperling (2003).

Climate change may also have much broader implications for development planning and development co-operation activities in a much longer time frame. Infrastructure, which is a critical vehicle for economic development, could be particularly sensitive to climate change impacts. Maintenance costs for long-lived assets like infrastructure typically have a "bathtub" curve, declining after an initial stabilisation period and beginning to increase again only after a long time as wear and tear take their toll towards the end of the asset's useful life. Meanwhile, changes in and variability of mean temperature, precipitation and sea levels are projected to increase progressively. Climate change effects on infrastructure will thus be more significant just when it is reaching the end of its useful life – a combination likely to increase the economic impact of climate change on infrastructure (Shukla, Kapshe and Garg, 2004: Figure 1.1). Moreover, equity implications can be expected: inundation of a highway or rail line might cut off critical access for those in the most remote areas or groups with no alternate mode of transport or communication.

Figure 1.1. **Climate change impacts on infrastructure maintenance costs**

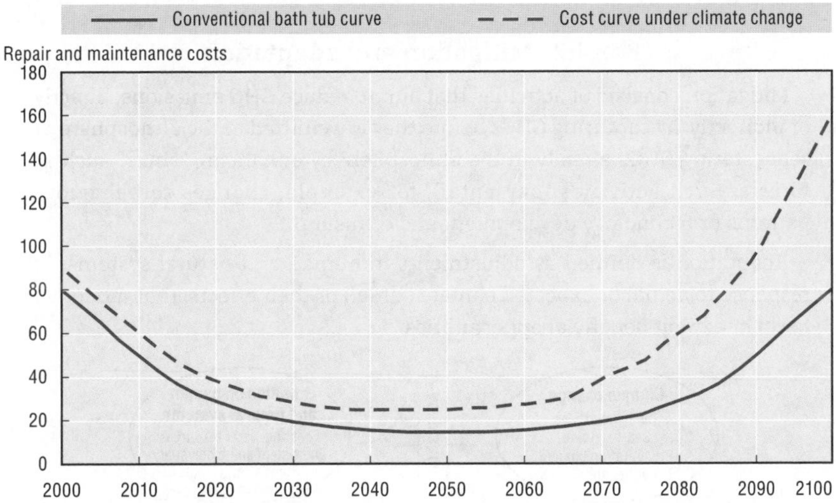

Source: Shukla, Kapshe and Garg (2004).

Thus, the impact of climate change is part of the larger question of how complex social, economic and environmental systems interact and shape prospects for development. There are multiple links. Economic development affects ecosystem balance and, in turn, is affected by the state of the ecosystem. Poverty can be both a result and a cause of environmental degradation. Material- and energy-intensive lifestyles and continued high levels of consumption of non-renewable resources, as well as rapid population growth, are unlikely to be consistent with sustainable development paths. Similarly, extreme socio-economic inequality within communities and between nations may undermine the social cohesion that would promote sustainability and make policy responses more effective. At the same time, socio-economic and technology policy decisions made for reasons unrelated to climate have significant implications for climate policy and climate change effects, as well as for other environmental issues. In addition, critical impact thresholds and vulnerability to climate change are directly related to environmental, social and economic conditions and institutional capacity (Munasinghe, 2002).

4. Responses to climate change and the "mainstreaming" challenge

Like the problem of climate change itself, many of the proposed responses are closely intertwined with development choices and pathways. There are two broad categories of responses to climate change: *mitigation* and *adaptation* (Box 1.2). Both seek to reduce or avoid damage. While mitigation

Box 1.2. **Mitigation and adaptation**

Mitigation consists of activities that aim to reduce GHG emissions, directly or indirectly, by capturing GHGs before they are emitted to the atmosphere or sequestering GHGs already in the atmosphere by enhancing "sinks" such as forests. Such activities may entail, for example, changes to behaviour patterns or technology development and diffusion.

Adaptation is defined as adjustments in human and natural systems, in response to actual or expected climate stimuli or their effects, that moderate harm or exploit beneficial opportunities.

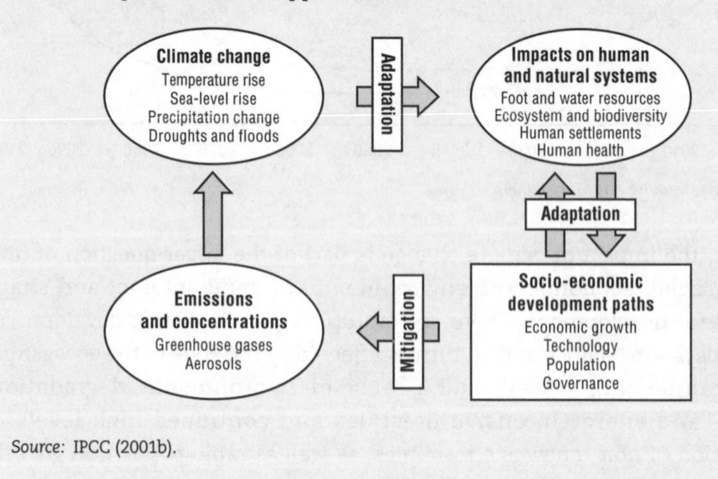

Source: IPCC (2001b).

BRIDGE OVER TROUBLED WATERS – ISBN 92-64-01275-3 – © OECD 2005

aims to reduce the causes of climate change by slowing the growth of GHG emissions, adaptation reduces the impact of climate stresses on human and natural systems. Adaptation strategies can be further classified as reactive or anticipatory, depending on when they are initiated. Both natural and human systems can adapt, but only human systems can undertake anticipatory adaptation. Within human systems, adaptation can be further classified in terms of whether the actions are undertaken by private or public agents.

Both mitigation and adaptation interact with development activities in a dynamic cycle, often characterised by significant delays. Mitigation and adaptation actions themselves can have implications on future development in the form of direct benefits of avoided climate damage on development prospects; ancillary benefits of mitigation and adaptation on other development activity; direct costs of mitigation or adaptation, which might impede development; and positive or negative spillover effects of climate policies in one location on other countries or regions, for example through trade and prices (Swart *et al.*, 2003).

Conversely, development levels and pathways strongly affect the capacity for both adaptation and mitigation. Development trends, as well as sectoral policies pursued for non-climate objectives, have the potential to significantly increase or decrease greenhouse emissions. For example, economic restructuring or a shift from manufacturing to services in an economy can lower emissions. Similarly, some policies pursued with development objectives, such as afforestation and promotion of biofuels, can be synergistic with mitigation objectives. On the other hand, development paths fuelled predominantly by continued or enhanced consumption of carbon-intensive energy sources will conflict with mitigation objectives. This situation presents both a challenge and an opportunity for many developing countries, which are projected to account for more than 60% of the increase in world primary energy demand, and roughly two-thirds of the increase in CO_2 emissions, over 2000-30 (IEA, 2002).

As regards adaptation, a range of development activities oriented towards reduced poverty and improved nutrition, education, infrastructure and health would automatically help decrease vulnerability to many climate change effects. A healthier, better-educated population with improved access to resources is also likely to be in a better position to cope with climate change; conversely, developing countries in general have a much lower "adaptive capacity" than their OECD counterparts (IPCC, 2001b). In many cases, however, business-as-usual development can promote maladaptation or exacerbate vulnerability to climate change, or both. Promoting human settlement or infrastructure development in areas that might become vulnerable to climate change effects is one example of development activities that could render a country maladapted. In other cases, however, certain development initiatives might *a priori* be "climate neutral" but nevertheless offer win-win opportunities to promote climate objectives if such considerations are taken into account in policy or project design and implementation.

Therefore, a two-track approach is needed in dealing with climate change. On the one hand it is important to pursue mitigation and adaptation through policy instruments intended specifically to deal with climate change. GHG mitigation has been the centrepiece of international efforts to combat climate change since the early 1990s, through the United Nations Framework Convention on Climate Change (1992, UNFCCC) and Kyoto Protocol (1997), the latter establishing near-term mitigation targets for industrialised countries. Although both agreements also contain elements of adaptation, that issue has emerged on an equal footing with mitigation in climate policy circles only since about 2001, as the following section will show.

On the other hand, given the critical links that mitigation and adaptation have with development, it is important to examine whether and how business-as-usual development can take such links into account. This is the rationale for "mainstreaming" climate change responses into regular development activity undertaken by various public and private agents. Mainstreaming offers the potential of significantly amplifying the ability to respond to climate change by involving a range of new actors: non-environment sections of national governments and international development agencies, as well as parts of the private sector and civil society that might otherwise not be engaged in the climate issue. It also offers the opportunity to better exploit the synergies and minimise the conflicts between development and climate objectives.

5. Status of adaptation policy and mainstreaming efforts

There is considerable information on how GHG mitigation can be mainstreamed into development priorities, particularly with regard to the roles of energy efficiency, renewable energy sources and afforestation in meeting dual climate change and development objectives – though implementation is a continued challenge. In the case of adaptation, however, fundamental issues remain as regards both successful implementation of stand-alone measures and efforts to mainstream them in development activity.

5.1. *Climate policy for adaptation*

Adaptation is a relatively recent concern in climate policy circles, and experience on adaptation practices and technologies is still limited. Adaptation is implicit in the ultimate objective of the UNFCCC (Article 2), which is to stabilise "greenhouse gas concentrations in the atmosphere at a level that would prevent dangerous anthropogenic interference with the climate system" and to achieve that level "within a time-frame sufficient to allow ecosystems to adapt naturally to climate change...". Even the extent to which human interference may be considered "dangerous" implicitly depends upon the ability of human and natural systems to adapt and thereby reduce the net damage.

At the first Conference of the Parties (COP-1), negotiators agreed on a three-stage approach for adaptation funding, overseen by the Global Environment Facility (GEF), the financial mechanism of the UNFCCC. Stage I focuses on climate change impact and vulnerability assessments as well as identification of adaptation responses; Stage II on implementation of measures (including capacity building) to prepare for adaptation; and Stage III on implementation of the adaptation measures themselves.

Adaptation is generally treated in a diffuse manner, however, without a clear policy objective (other than reporting requirements on actions undertaken) in the UNFCCC and Kyoto Protocol texts (Box 1.3).

The implementation of adaptation has come into much sharper focus since the negotiation of the Kyoto Protocol and, in particular, since COP-7 in Marrakech (2001) established three funds dealing with adaptation:

- The Least Developed Countries Fund is intended to address the particularly low adaptive capacity of the least developed countries (LDCs). The Marrakech Accords established the fund to help such countries prepare their National Adaptation Programmes of Action (NAPAs), which establish and prioritise adaptation needs. The fund also supports institutional capacity building and other activities.
- The Special Climate Change Fund finances a multitude of activities in both mitigation and adaptation in all developing countries. The activities can be specific to sectors – energy, transport, industry, agriculture, forestry and waste management – or aimed directly at adaptation, technology transfer and economic diversification.
- The Adaptation Fund – the only Marrakech fund linked to the protocol rather than the convention – provides funding only to Parties to the protocol. Like the other two funds, its resources come from voluntary contributions, but it also benefits from a 2% share of the proceeds of certified emission reductions from projects under the protocol's Clean Development Mechanism.

The Marrakech funds are part of a more complex architecture of international funding sources for adaptation (Figure 1.2), including the GEF Trust Fund and bilateral initiatives by development agencies and non-governmental organisations (NGOs). The Adaptation Fund became operational only upon the entry into force of the Kyoto Protocol in February 2005. The LDC Fund, the most advanced of the three, has received around USD 20 million to support NAPAs. The Special Climate Change Fund became operational as of COP-9 in late 2003, but has not received any contributions yet. The GEF recently initiated a USD 50 million

Box 1.3. **UNFCCC and Kyoto Protocol articles in support of adaptation**

UNFCCC

Article 4.1(b): All Parties shall "Formulate, implement, publish and regularly update national and, where appropriate, regional programmes containing measures to mitigate climate change by addressing anthropogenic emissions by sources and removals by sinks of all greenhouse gases not controlled by the Montreal Protocol, and measures to facilitate adequate adaptation to climate change".

Article 4.1(e): All Parties shall "Co-operate in preparing for adaptation to the impacts of climate change; develop and elaborate appropriate and integrated plans for coastal zone management, water resources and agriculture, and for the protection and rehabilitation of areas, particularly in Africa, affected by drought and desertification, as well as floods".

Article 4.1(f): All Parties shall "Take climate change considerations into account, to the extent feasible, in their relevant social, economic and environmental policies and actions, and employ appropriate methods, for example impact assessments, formulated and determined nationally, with a view to minimizing adverse effects on the economy, on public health and on the quality of the environment, of projects or measures undertaken by them to mitigate or adapt to climate change".

Article 4.4: "The developed country Parties and other developed Parties included in Annex II shall also assist the developing country Parties that are particularly vulnerable to the adverse effects of climate change in meeting costs of adaptation to those adverse effects."

Kyoto Protocol

Article 10(b): Each Party shall "Formulate, implement, publish and regularly update national and, where appropriate, regional programmes containing measures to mitigate climate change and measures to facilitate adequate adaptation to climate change."

Article 10(b)(i): "Such programmes would, *inter alia*, concern the energy, transport and industry sectors as well as agriculture, forestry and waste management. Furthermore, adaptation technologies and methods for improving spatial planning would improve adaptation to climate change."

Article 10(b)(ii): Parties shall submit information on action under the Protocol, including adaptation measures.

Article 12.8: "The Conference of Parties… shall ensure that a share of the proceeds from certified project activities is used… to assist developing country Parties that are particularly vulnerable to the adverse effects of climate change to meet the costs of adaptation."

BRIDGE OVER TROUBLED WATERS – ISBN 92-64-01275-3 – © OECD 2005

Figure 1.2. **International architecture for adaptation funding**

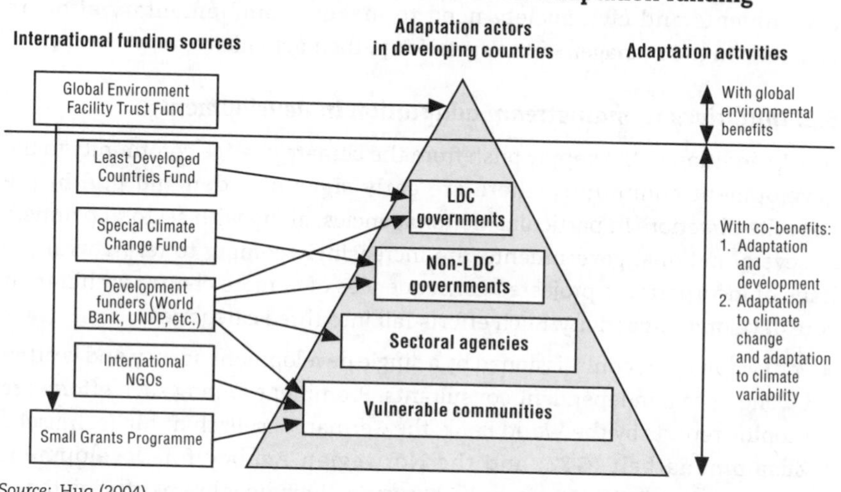

Source: Huq (2004).

pilot adaptation action programme; guidelines for disbursement are still under negotiation.

Considerable progress has thus been made in terms of prioritising and funding adaptation within the UNFCCC process. Addressing adaptation entirely through multilateral climate negotiations may have certain limitations, however, and the voluntary nature of the Marrakech funds could affect their predictability over the long term. Moreover, the requirement that projects must demonstrate "global environmental benefits" to receive grants from the GEF Trust Fund (except in the Small Grants Programme) has constituted a critical challenge for adaptation projects. Such benefits are automatic in the case of mitigation, as emission reductions anywhere help reduce the problem at global level. The benefits of adaptation, however, tend to accrue primarily to regions, communities and individuals who undertake such actions. Even spillover effects are generally restricted to local or regional level. In addition, GEF funds are intended to meet "full agreed incremental costs of implementing measures" to address anthropogenic climate change, but because adaptation activities respond to a continuum of climate stresses, from natural variability to longer-term human-induced climate change, delineating increments is difficult.

Beyond such specific issues is the concern that the climate-negotiations approach encourages "global managerialism" in addressing adaptation, with often slow, top-down, globally negotiated outcomes (Adger *et al.*, 2003). Further, despite the magnitude of financial commitments to international adaptation activities, such efforts are probably dwarfed by the magnitude of investment in national and local development activities that are potentially vulnerable to climate risk (including that of climate change). Thus, development agencies, governments and civil society need to make a complementary effort to mainstream the management of such risk in their activities.

5.2. *Initiatives to mainstream adaptation in development*

In addition to the supply push from the climate change community to the development community, there are early signs of a demand pull in the opposite direction. In particular, donor agencies, along with sectoral planners in several national governments, are increasingly coming to terms with the issue of incorporating projected consequences of climate change within their core development activity. Such efforts fall into three categories:

- *Portfolio analysis*, commissioned by a single development agency and written by one or two independent consultants. Prominent among such efforts are scoping reports by the World Bank, the German Gesellschaft für Technische Zusammenarbeit (GTZ) and the Norwegian Agency for Development Co-operation (NORAD) on links between their development assistance portfolios and climate change adaptation. Burton and van Aalst (1999), for instance, conclude that a significant proportion of World Bank investment is at risk from direct and indirect effects of climate change and variability. The study provides examples from South America, Asia and the Pacific where climate risk assessments were not considered during project preparation. Noting the lack of attention to climate change in procedures regarding project design, implementation and evaluation, the authors advise conducting further studies, providing training and undertaking institutional changes to raise awareness.

  Another example is Klein (2001), analysing whether adaptation to climate change is mainstreamed into German Official Development Assistance (ODA). The author reviewed descriptions of 136 German-funded projects, which focused on natural resource management in Africa. None referred to climate change and only a few mentioned environmental or economic stresses related to weather or climate variability, such as drought and desertification; instead, the emphasis was on more immediate issues, such as food security, sanitation, safe water supply and education.

  The most recent such analysis, by Eriksen and Næss (2003), examines Norwegian development co-operation activities relevant to climate change adaptation, with a special focus on links with poverty reduction strategies in rural areas in Africa. The analysis is based on a review of policies and strategies, not on particular development projects. The report identifies three strategic entry points (livelihood strategies, local capacities and sensitivity, and risk management and early warning systems), which could be used in development co-operation activities as foci for adaptation support.

- *Comprehensive country or regional assessments* involving large teams of national and international experts, as well as consultation with government and other stakeholders. The two major initiatives in this category were initiated by the World Bank, in the Pacific Islands and Bangladesh. *Cities, Seas, and Storms*, assessing how Pacific Island economies will deal with changes in urbanisation, use of ocean resources, and climate, concludes that climate change would have a significant impact on most Pacific Islanders and that immediate development choices would profoundly affect future vulnerability to climate change. Given the uncertainties in climate change projection, the report primarily recommends such "no-regrets" adaptation measures as improved management of natural resources and the mainstreaming of adaptation goals in expenditure planning and policy development for a wide range of sectors (World Bank, 2000b).

  The Bangladesh study identified physical and institutional adaptation options and suggested a strategic, cross-cutting adaptation approach involving co-ordinated institutional response at national level to enhance integrated planning, research and dissemination, as well as increased international activities in areas such as cross-boundary river flows (World Bank, 2000a). Results from this study appear to have been embraced in some areas, including coastal zone management programmes, cyclone preparedness plans and a new 25-year water sector plan (Rahman and Alam, 2003). The report was less successful in convincing high-level policy makers and central bodies such as the Finance Ministry of the importance of taking climate change into account as part of development planning.
- *High-level policy documents*, which constitute a relatively new category of mainstreaming initiatives. Instead of project or country-level analyses, this category focuses on high-level policy statements on the need for mainstreaming climate change concerns in development co-operation. In 2003, ten multilateral and bilateral organisations, including the OECD, affirmed the central importance of climate change effects and adaptation for the achievement of poverty alleviation goals (Sperling, 2003). While the document primarily served only an awareness-raising function with no direct follow-up, its endorsement by senior officials from the participating agencies indicated the growing significance of climate change for the development agenda.

  Another high-level initiative with potentially more direct programmatic implications is under way at the European Commission, which adopted a climate change strategy for support to partner countries in March 2003 (European Commission, 2003). Recognising that developing countries are the most vulnerable and that the impact of climate change is likely to impede their development, the EU is seeking to assist partner countries to meet these challenges through mainstreaming of climate concerns in EU development co-operation. The strategy was translated into an action plan, which emphasises adaptation as well as capacity development and research: in November 2004, the General Affairs and External Relations Council adopted the "Action Plan to accompany the EU Strategy on Climate Change in the Context of Development Co-operation" for 2004-08 (see Chapter 5, Box 5.1).

These high-level policy initiatives are not specific to particular country contexts, and it is too early to assess their operational impact on in-country development co-operation policies.

This growing interest from the development community on integrating climate change concerns, particularly adaptation, into initiatives has largely been prompted and co-ordinated by development agencies' climate experts. Such efforts have generally served an awareness-raising function. Other than the Bangladesh and Pacific Islands studies, they have generally not included explicit consideration of how climate change projections (and associated uncertainties) in specific countries and regions affect development activity. As for project analyses, thus far they have been limited to selected activities in a handful of countries undertaken by single donors. Moreover, despite a flurry of interest in climate change adaptation from parts of the development community, others express a fair degree of scepticism. How does a global concern like climate change, which might manifest itself decades to centuries in the future, compare with more immediate, visible, local and regional priorities for development, such as poverty, food security, sanitation and public health? How can one assess the exposure of development portfolios to potential risk related to climate change? In any case, coping with the impact of weather extremes is already an integral part of many development activities, such as famine early warning and flood plain management; would climate change really require anything different in operational terms?

Such questions indicate that further analyses are needed to examine the relative significance of implications of climate change in particular settings. There is also a need to establish frameworks for prioritising action on climate change given the uncertainties in projections, assess exposure of development investments in activities potentially vulnerable to climate risk, and examine synergies and conflicts between climate responses and development priorities and plans in specific contexts.

6. Objectives and scope of this study

The overall objective of this study is to explore synergies and trade-offs in mainstreaming responses to climate change within development planning and assistance, with natural resource management as an overarching theme. There is a particular emphasis on implications for development co-operation activities of OECD donors, as well as for national and regional planners in developing countries.

The primary focus is on mainstreaming adaptation to climate change, though links between development objectives, natural resource management policies and GHG mitigation are also examined. Adaptation can be viewed on three levels: *i)* adaptation to current variability; *ii)* adaptation to observed medium- and long-term climate trends; and *iii)* anticipatory planning in response to model-based scenarios of long-term climate change (Figure 1.3). Responses at the three levels may be linked to each other and to other considerations and development priorities. Adapting to current climate fluctuations is sensible in terms of economic development, given direct and certain evidence of the adverse effects involved. Such adaptation is also likely to enhance societies' resilience, enabling them to cope with further consequences of climate change, since many human-induced changes in climate will manifest themselves through enhanced or altered climate variability (Agrawala and Cane, 2002). Nevertheless, without adequate analysis it would be premature to rule out the possibility that anthropogenic climate change may also require forward-looking investment and planning responses that go beyond short-term responses to current climate variability. This possibility is becoming increasingly important, given the growing evidence of detectable long-term trends in climate patterns and observed associated effects, as the IPPC Third Assessment Report concludes (IPCC, 2001a and b). Also relevant are routine investment and infrastructure decisions, which leave a footprint for several decades or more and might therefore need to incorporate scenarios of future climate change.

Figure 1.3. **Levels of adaptation responses and links to other priorities**

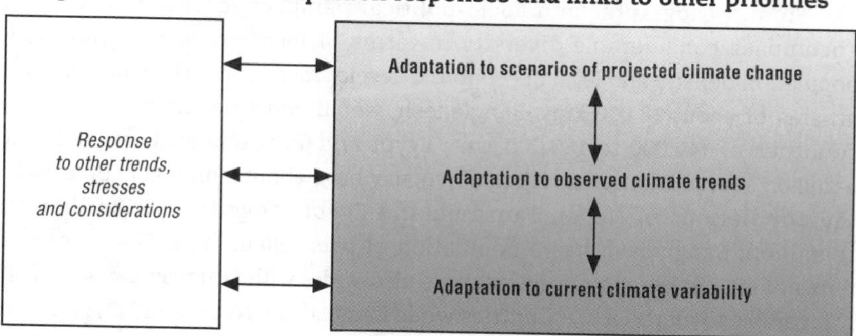

Thus, the study takes as its focus the links between economic development planning and climate responses to climate variability and change over the medium term, from several years to a few decades. It includes development policies and projects that have a "locked-in" character, in that they might enhance or constrain societies' ability to cope with climate variability and change over the medium term. It also includes consideration of new planning responses that may be required to cope with the impact of climate changes that might manifest themselves in coming years.

6.1. *Case studies: unit of analysis, selection and overview*

The overall unit of analysis in this study is at the national level, although analysis of particular responses will have a sectoral or regional/local focus. Climate change impact and vulnerability do not follow political boundaries, but economic planning, development assistance, and mitigation and adaptation responses often do. Country case studies were conducted in Bangladesh, Egypt, Fiji, Nepal, Tanzania and Uruguay. The selection of countries was based upon climate-related and socio-political considerations.

The six countries present a range of climates, from tropical and oceanic to high mountain, desert and temperate. Each country is already experiencing the impact of current climate variability in the form of phenomena such as flooding, drought, coastal inundation, cyclones and storm surges. Some countries are also experiencing the effects of long-term trends in climate variables, most notably glacier retreat and glacial lake expansion in Nepal. There are also indications that the case study countries might be adversely affected by long-term climate change through sea level rise, temperature increases and changes in rainfall, along with associated consequences such as coastal inundation, loss of wetlands and farmland, melting of permafrost and tropical glaciers, and reduced water use efficiency. Figure 1.4 shows the case study countries and the focus areas for in-depth analysis.

From a geographic and socio-political perspective, the six countries encompass considerable diversity in terms of location, size, population, population density and level of economic development. Fiji is the smallest with an area of about 18 000 km^2; Bangladesh, Nepal and Uruguay are mid-sized countries of 140 000 to 180 000 km^2; Egypt and Tanzania each cover about a million square kilometres. Fiji and Uruguay have about a million people each; the populations of Nepal, Tanzania and Egypt range from 25 million to 70 million; Bangladesh has a population of over 130 million. The spread in terms of overall human development is also wide, with Uruguay being almost comparable to many OECD countries while Bangladesh, Nepal and Tanzania are ranked in the lowest tier of countries on the UNDP Human Development Index.

Collectively, the case studies make it possible to examine the opportunities and challenges presented by implementation and mainstreaming of climate change objectives in a wide variety of geographic and socio-political contexts. While this presentation lends itself to some cross-cutting findings and patterns, obviously the degree to which one can generalise from many of the conclusions is limited, given the small number of case studies.

BRIDGE OVER TROUBLED WATERS – ISBN 92-64-01275-3 – © OECD 2005

Figure 1.4. **Case study countries with focus areas for in-depth analysis**

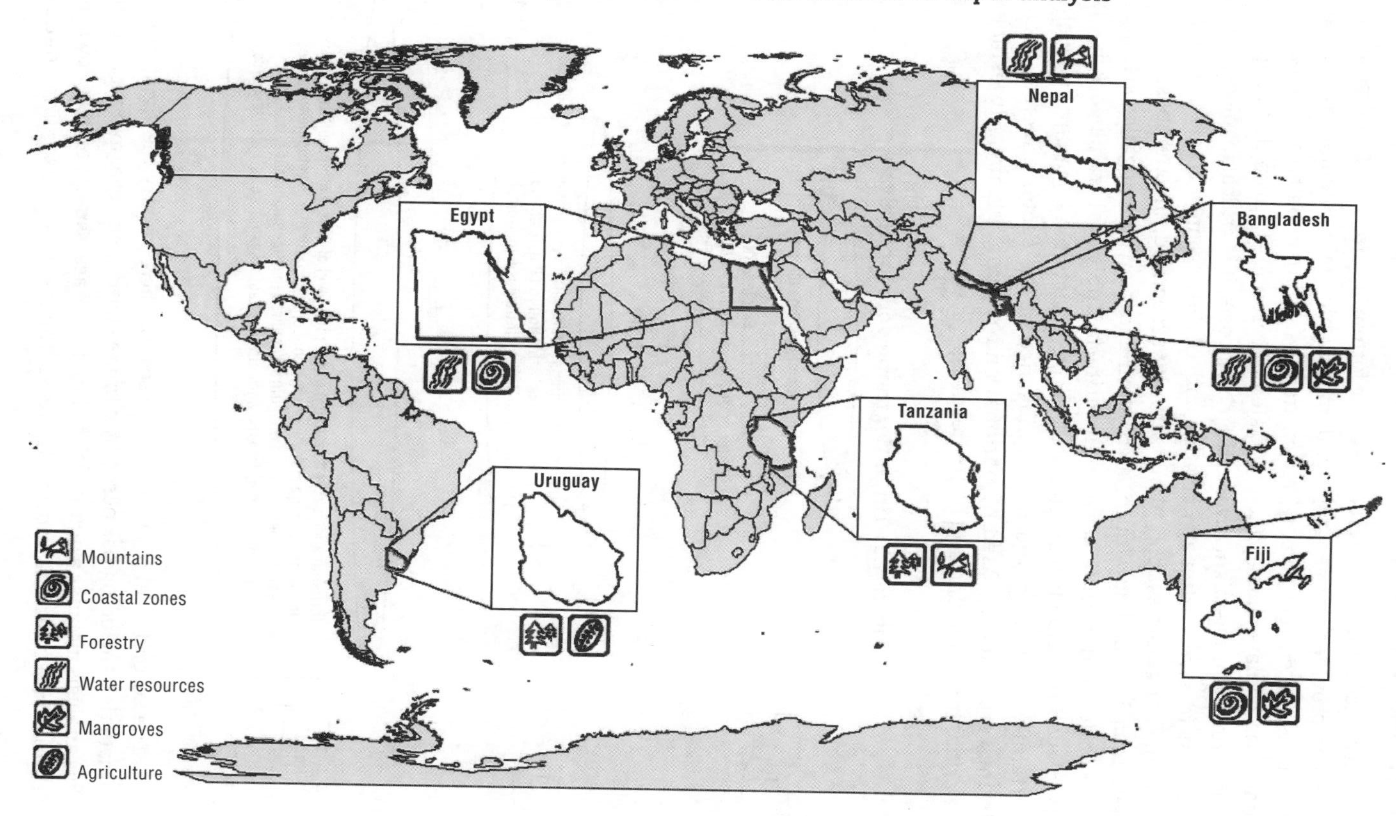

7. Framework of analysis

Each case study follows a three-tier format (Figure 1.5). The first level of analysis provides the geographic, demographic and economic context and examines the current climate, recent trends in climate patterns, scenarios of long-term climate change and associated uncertainties. Given that adaptation and mainstreaming might involve their own costs, it is critical for development planners to have a better appreciation both of the baseline climate risks and of climate change projections, along with the associated uncertainties, so as to determine how, and how much, such information can meaningfully be factored into policies and projects. A multi-criteria analytical framework is also developed to help classify climate change effects based upon the level of certainty, the timing and the significance of the sector affected. Such a framework could be used by national and regional planners to establish priorities for adaptation and mainstreaming.

Figure 1.5. **Three-tier framework for analysis**

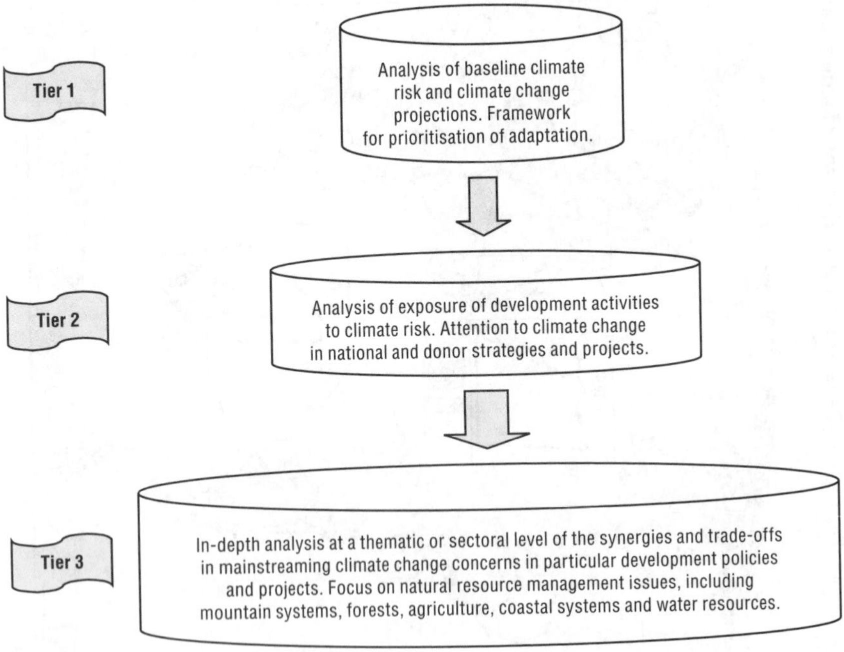

Tier 1

Analysis of baseline climate risk and climate change projections. Framework for prioritisation of adaptation.

Tier 2

Analysis of exposure of development activities to climate risk. Attention to climate change in national and donor strategies and projects.

Tier 3

In-depth analysis at a thematic or sectoral level of the synergies and trade-offs in mainstreaming climate change concerns in particular development policies and projects. Focus on natural resource management issues, including mountain systems, forests, agriculture, coastal systems and water resources.

The second level of analysis assesses for each country the significance of climate risk, including risk posed by climate change, on development activities. This analysis includes a quantitative climate risk assessment of ODA flows from all principal bilateral and multilateral donors. It also examines national planning documents, such as national development plans, poverty reduction strategy papers and sectoral development strategies, as well as donor agencies' country assistance and sectoral strategies, so as to assess the degree of attention to climate risk (and climate change in particular) in such documents. Some specific development projects in particularly vulnerable sectors are also examined for their degree of attention to climate change.

The third level takes an in-depth look at particular natural resource management areas that might be particularly vulnerable to the effects of climate change or critical from the perspective of GHG mitigation. The intent is to analyse in detail the synergies and trade-offs between climate change objectives and sectoral or development priorities. The systems examined cover natural resource management themes ranging from mountain systems, forests and agriculture to coastal systems and water resources.

8. Outline of this volume

The individual, integrated case studies for Bangladesh, Egypt, Fiji, Nepal, Tanzania and Uruguay, following the three-tier analytical framework, have been published separately (OECD, 2003a-d and 2004a-b). This synthesis volume summarises key results from the individual case studies and highlights patterns and cross-cutting results across the six studies. Chapter 2 discusses current climate as well as climate change projections for the six countries. It gives particular attention to the quantification of uncertainties in climate change projections which might condition the urgency and extent of adaptation responses. Chapter 3 assesses the degree of exposure of development co-operation activities to climate risk, and analyses key national plans, donor strategies and project documents for their attention to climate change. Chapter 4 assesses in depth the opportunities and constraints involved in linking climate change and sectoral objectives in several key areas of natural resource management, including water resource management in the Himalayas and the Nile system, ecosystem protection on Mount Kilimanjaro and conservation of coastal mangroves in Bangladesh and Fiji. Chapter 5 concludes the volume with a review of key findings.

ISBN 92-64-01275-3
Bridge Over Troubled Waters
Linking Climate Change and Development

Chapter 2

Climate Analysis

by
Joel B. Smith, Shardul Agrawala, Peter Larsen and Frédéric Gagnon-Lebrun

Development planners need a better appreciation of climate change projections and the associated uncertainty in order to determine how to factor this information into their activities, and to what extent. Full understanding might ultimately require climate change risk assessment tailored to the locations, variables and time horizons of relevance to specific development activities, but that is beyond the scope of this analysis. This chapter provides an overview of the baseline climate for each case study country, followed by climate change projections along with an assessment of the associated uncertainty. A multi-criteria analytical framework is then proposed to classify climate change impacts by level of certainty, timing and the significance of the sector affected. National and regional planners could use such a framework to establish priorities for adaptation and mainstreaming.

1. Introduction

The extent to which particular development activities need to take climate change impacts into account depends upon the degree to which the impacts might manifest themselves over the planning horizons of the activities. Moreover, a specific development project can involve different planning horizons for different stakeholders. For example, in the case of water resource or catchment-based assessment, planning for water storage might be for 50 years or more, while that for water supply might be for 5 to 15 years (Jones *et al.*, 2004). Incorporation of long-term climate risks could thus be more important when planning for water storage than for water supply. Figure 2.1 shows typical time horizons for selected initiatives.

On the other hand, development policies (as opposed to projects) often have short- to medium-term horizons, for example as part of multi-year

Figure 2.1. **Representative time horizons (in years) for climate risk assessments**

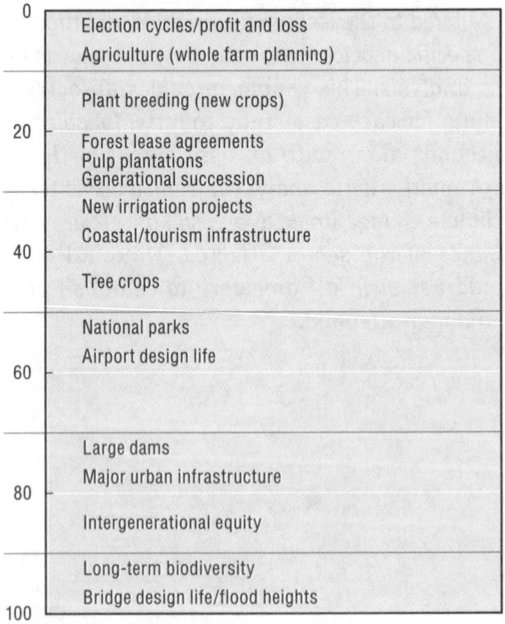

0
Election cycles/profit and loss
Agriculture (whole farm planning)

Plant breeding (new crops)
20
Forest lease agreements
Pulp plantations
Generational succession

New irrigation projects
Coastal/tourism infrastructure
40
Tree crops

National parks
Airport design life
60

Large dams
Major urban infrastructure
80
Intergenerational equity

Long-term biodiversity
Bridge design life/flood heights
100

Source: Jones *et al.* (2004).

national or sectoral plans. Opportunities to update or otherwise modify policies may therefore arise every few years, or at least within a decade or two. Thus, a project such as a coastal railway with a lifetime of several decades to a century might need to incorporate century-scale projections of sea level rise, whereas policies on coastal defence and early warning could be updated at relatively short notice in response to changing trends in climate risk, at least theoretically. In fact, however, policies often engender irreversible outcomes over much longer time horizons. For example, coastal zoning can spur the development of settlements, roads and infrastructure. It would be financially and politically expensive to reverse these even if the area were later to become much more vulnerable to sea level rise. Policies that could have this type of long-term "footprint" might need to take into account the implications of climate change over several decades or more as well.

The time horizon is only one screening criterion for projects and plans where climate change and its impacts might be relevant. The specificity and quality of climate change and impact projections determine how, and to what extent, such information can be productively incorporated in development planning. There is often an inverse relationship between quality and the spatial specificity of climate change projections. Global averages, such as those presented in IPCC assessments, tend to be more robust than location-specific projections but also less relevant for decision makers and planners, who operate at local to national scale. The level of uncertainty associated with projections also depends upon the variable involved. Projections of temperature are often more robust than those of precipitation, for instance, and projections of second- or higher-order impacts (*e.g.* on particular crops) are more uncertain than those of first-order biophysical variables such as temperature and sea-level rise.

Given that adaptation and its mainstreaming in development activities may involve their own costs, it is critical for development planners to have a better appreciation of baseline climate risks, climate change projections and the associated uncertainty. Such understanding will help them determine how to factor the information into their policies and projects, and to what extent. Full understanding might ultimately require climate change risk assessment that is tailored to the particular location, climate variables and time horizons of relevance to specific development policies or projects, but that is beyond the scope of this analysis. This chapter provides an overview of the baseline climate for each of the six case study countries, followed by discussion of how to quantify the uncertainty associated with climate change projections, as this determines the reliability of projections to be used as a basis for adaptation planning. A multi-criteria analytical framework is then proposed, to help classify climate change impacts by level of certainty, timing and the significance of the sector affected. Such a framework could be used by national and regional planners to establish priorities for adaptation and mainstreaming.

2. Baseline climate

The case study countries – Bangladesh, Egypt, Fiji, Nepal, Tanzania and Uruguay – vary considerably in location, size and topography, all of which condition their respective climatic circumstances.

Bangladesh has a humid, tropical climate that is influenced by the monsoon and, to some extent, by the pre- and post-monsoon circulation. The four main seasons are winter (December to February), pre-monsoon (March to May), monsoon (June to early October) and post-monsoon (late October to November). The winter is cooler and drier, with average temperatures ranging from 7.2-12.8 °C to 23.9-31.1 °C. Generally the south is about 5 °C warmer than the north. The pre-monsoon is hot, with an average temperature of 36.7 °C and peaks of close to 41 °C in certain regions. The monsoon is hot and humid, with torrential rainfall throughout the season. The short post-monsoon season is characterised by withdrawal of the monsoon and a gradual lowering of night-time temperatures. The mean annual rainfall is about 2 300 mm, but the geographic and seasonal variations are wide; annual rainfall ranges from 1 200 mm in the extreme west to over 5 000 mm in the east and north-east.

Egypt's climate is semi-desert, characterised by hot summers, moderate winters and very little rainfall. Strong winds occur in some areas, especially along the Red Sea and Mediterranean coasts: sites with annual average wind speed of 8.0-10.0 metres per second have been identified along the Red Sea coast and of about 6.0-6.5 m/s along the Mediterranean coast. Average precipitation in the Ethiopian highlands, where much of the water of the Nile originates, is highest in July, August and September, at 5.4 mm per day, and almost negligible between January and March.

Fiji has an oceanic tropical climate, with seasonal and inter-annual variations. Temperatures range from 23 °C to 25 °C during the dry season (May to October) and from 26 °C to 27 °C during the rainy season (November to April). Rainfall distribution is strongly influenced by terrain: leeward sides of mountainous islands tend to be drier and windward sides tend to be wetter. On the island of Viti Levu, for example, rainfall ranges from 3 000 mm to 5 000 mm on the windward side and from 2 000 mm to 3 000 mm on the leeward side. Cyclones are a major weather concern: the highest concentration of cyclones in the South Pacific occurs in Fiji's waters. Cyclones can have a major economic and public safety impact, causing up to 25 deaths and damage valued at FJD 170 million (about USD 85 million) in one event (Feresi *et al.*, 1999). Periodic droughts are another concern. El Niño events generally position the South Pacific Convergence Zone north-east of the island and result in hotter, drier conditions from December to February and cooler, drier conditions from June to August.

Nepal contains eight of the world's ten highest mountain peaks, including Mount Everest (8 848 metres), as well as areas lying only about 80 metres above

BRIDGE OVER TROUBLED WATERS – ISBN 92-64-01275-3 – © OECD 2005

sea level. Such altitudinal differences result in extreme spatial variation in climate, from tropical to arctic within only about 200 km (the size of an average grid box in a climate model). Much of Nepal is in the monsoon region, and climate variation within the country is largely a function of elevation. The mean temperature is around 15 °C, with higher temperatures in the south and in mountain valleys. Average rainfall is 1 500 mm, with rainfall increasing from west to east. North-western Nepal, situated in the rain shadow of the Himalayas, has the least precipitation. Rainfall also varies by altitude, with areas over 3 000 metres receiving a lot of drizzle while heavy downpours are common below 2 000 metres. Although the annual average rainfall is high, seasonal distribution is of great concern: flooding is frequent in the monsoon season, during the summer, while drought is not uncommon in certain regions in other parts of the year.

Tanzania's climate ranges from tropical to temperate (in the highlands). Average annual rainfall is 1 042 mm, but only about half the country receives more than 762 mm annually. Average temperatures range between 17 °C and 27 °C, depending on location. Natural hazards include both flooding and drought. Altitude plays a large role in determining rainfall patterns in the country, with higher elevations receiving more precipitation. Tanzania has two rainfall regimes: bimodal, with long rains from March to May and short rains from October to December, in much of the north, and unimodal, with most rainfall from December to April, in most of the rest of the country.

Uruguay is located entirely within the temperate zone and has a fairly uniform climate. Seasonal variations are pronounced, but extremes in temperature are rare. Precipitation is fairly evenly distributed throughout the year. Annual rainfall increases from south-east to north-west. Montevideo averages 950 mm annually. High winds are common during the winter and spring, and wind shifts are sudden and pronounced. In summer, winds off the ocean temper the warm daytime temperatures (Hudson and Meditz, 1990). Natural hazards are mainly linked to climate events.

3. Climate change projections

While there is a fair amount of information on regional and national climate change projections in documents such as national communications under the UNFCCC, it is often based on point projections from a limited number of general circulation models (GCMs). Such information does not give decision makers an appreciation of the robustness of climate projections, particularly whether projections from different climate models are similar or widely divergent. Comparison among models is better accomplished through the MAGICC/SCENGEN climate scenario generator, which converts GHG emission scenarios into estimates of future global temperature and sea level change, and then into descriptions of changes in regional climate (Box 2.1 and Figure 2.2).

Box 2.1. A brief description of MAGICC/SCENGEN

The MAGICC/SCENGEN software package enables investigations of future climate change based on emission scenarios for GHGs and other gases. Although MAGICC and SCENGEN are stand-alone programme packages, when they are used in conjunction, SCENGEN exploits MAGICC's output to develop climate and climate change scenarios at both global (mean) and regional levels and to assess related uncertainty.

MAGICC (Model for the Assessment of Greenhouse gas Induced Climate Change) consists of software that estimates annual mean global surface air temperature and sea level rise for particular emission scenarios and determines the sensitivity of these estimates to changes in the model parameters. It thus is a tool for comparing the global implications of scenarios, which may be generated for any period between 1990 and 2100 and for any combination of emissions of GHGs (CO_2, CH_4, N_2O, HFCs, PFCs, SF_6) as well as CO, SO_2, NO_x and VOCs (Raper *et al.*, 1996). The estimates are computed using a set of linked simple gas cycle models and climate models that, collectively, emulate the behaviour of fully three-dimensional, dynamic GCMs. The IPCC has used MAGICC as its primary model to produce projections. MAGICC employs IPCC Third Assessment Report science (see IPCC, 2001a).

SCENGEN (Global and Regional Climate SCENario GENerator) is a regionalisation algorithm using a scaling method developed by Santer *et al.* (1990) that constructs a range of spatially detailed climate change scenarios on a 5° latitude/longitude grid. SCENGEN generates regional scenarios for changes in or absolute values of temperature and precipitation, changes in or absolute values of temperature and precipitation variability, signal-to-noise ratios based on inter-model differences or temporal variability, and probabilities of temperature and precipitation change above a specified threshold. The algorithm exploits three sources of data: the output from MAGICC, results from a large number of GCM experiments and a data set of observed global and regional climate trends from 1961 to 1990. The GCM set examines patterns of regional climate change induced by GHG emissions through 16 coupled atmosphere-ocean GCMs and by sulphate aerosol emissions through the Urbana-Champaign GCM. For four regions – Europe, South Asia, the United States and southern Africa – SCENGEN contains observed climate data at 0.5° latitude/longitude resolution. SCENGEN rests on two main assumptions, which may be reasonable approximations for many regions and for some climate variables: *i)* the relative geographic patterns of anthropogenic change in climate as averaged over 30 years will remain constant over time; and *ii)* the magnitude of the changes is well described by the global mean temperature change related to each forcing component.

Source: Hulme *et al.* (2000).

BRIDGE OVER TROUBLED WATERS – ISBN 92-64-01275-3 – © OECD 2005

Figure 2.2. **Schematic of MAGICC/SCENGEN**

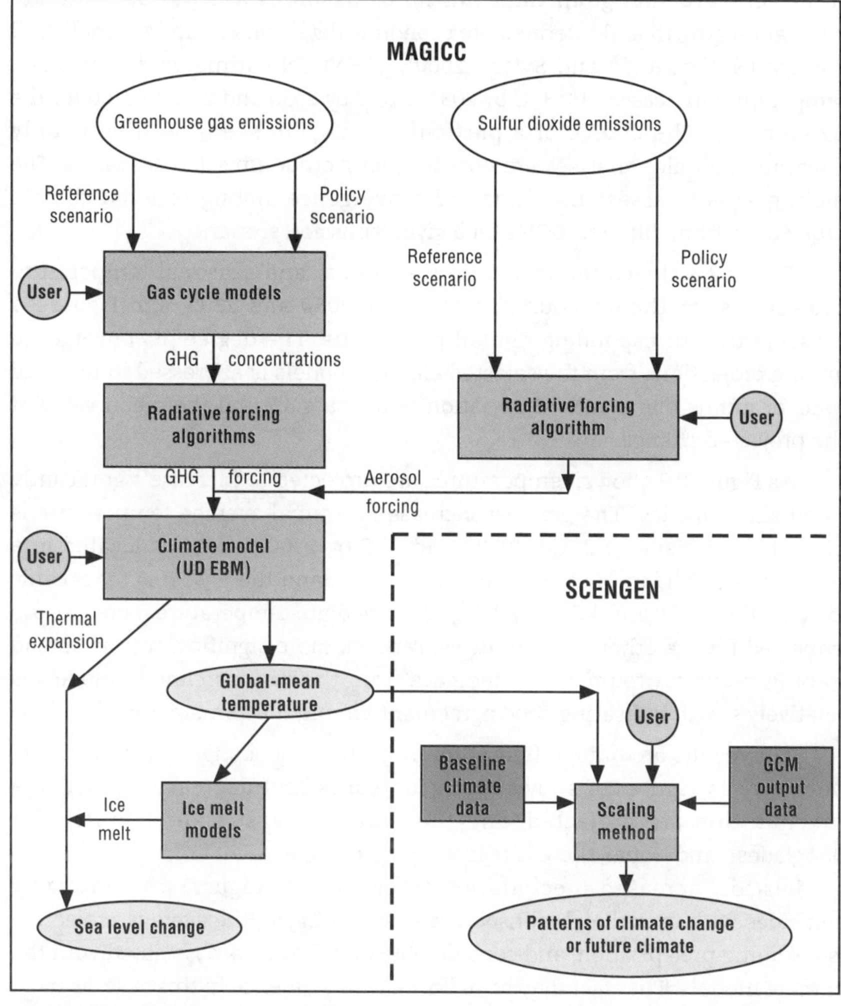

Source: Hulme et al. (2000).

3.1. *Temperature and precipitation*

The analysis presented here estimates changes in temperature and precipitation in each case study country by using MAGICC/SCENGEN to assess the associated uncertainty. Smith *et al.* (2003) conducted the analysis, whose results are discussed in the individual country case studies (OECD, 2003a-d and OECD, 2004a-b). First, simulations from 17 GCMs made since 1995 (Annex 2.A1) were examined to assess which models best estimate current precipitation for the country concerned. Next, one-half to two-thirds of the models retained were driven with the IPCC SRES B2 emission scenario to generate scenarios for future climate change in the country. B2 assumes a world of moderate population growth and intermediate economic development and technological change (Nakicenovic and Swart, 2000). SCENGEN estimates global mean temperature increases of 0.8 °C by 2030, 1.2 °C by 2050 and 2 °C by 2100 for the B2 scenario. The choice of a particular emission scenario here is only illustrative; similar analyses are possible using other emission scenarios. The main goal is to assess the degree of convergence among regional climate projections from different GCMs for a given emission scenario.

Figure 2.3 shows the results of the annual and seasonal temperature projections for the six countries for 2030, 2050 and 2100, and Figure 2.4 presents the corresponding rainfall projections. The degree of convergence among projections from the selected climate models is expressed in terms of (plus or minus one standard deviation) error bars around the mean value of the projected change.

As Figure 2.3 shows, temperatures are projected to increase significantly in all six countries. The greatest increase in annual average temperature is projected for Nepal (1.2 °C by 2030 and 3 °C by 2100), which is located in a mountainous region in the continental interior, and the lowest is for oceanic Fiji (0.6 °C by 2030 and 1.5 °C by 2100). Intermediate temperature increases are projected for the other four countries. Perhaps more significantly, the spread in projections for temperature increase from the various climate models is relatively small, indicating good agreement among the climate models.

The results are quite different for projections of change in precipitation, however, as Figure 2.4 shows. Both increases and decreases in average precipitation are projected, and they vary by season and country. For Bangladesh and Nepal, there is reasonably good agreement among the climate models for increased precipitation from June to August, which roughly coincides with the critical monsoon season. For Egypt, the average projection is declining precipitation, and considerable uncertainty is associated with the climate models. That fact may be of limited consequence in this case, as most of Egypt's agriculture is irrigated and the country's water demand is met by Nile flows originating outside its borders. Nile flows are sensitive to precipitation in headwaters areas such as the Ethiopian highlands, and to temperatures in Sudan and Egypt, which determine evaporation loss.

In general, the spread in precipitation projections from the various climate models is quite wide for most countries. Given this degree of uncertainty, in many cases it is difficult to assess whether precipitation would increase, decrease or remain more or less unaltered under climate change. Moreover, the uncertainty associated with climate change impacts driven primarily by changes in precipitation (*e.g.* those affecting rain-fed agriculture) might also be

Figure 2.3. **Projections for country-average temperature increase**

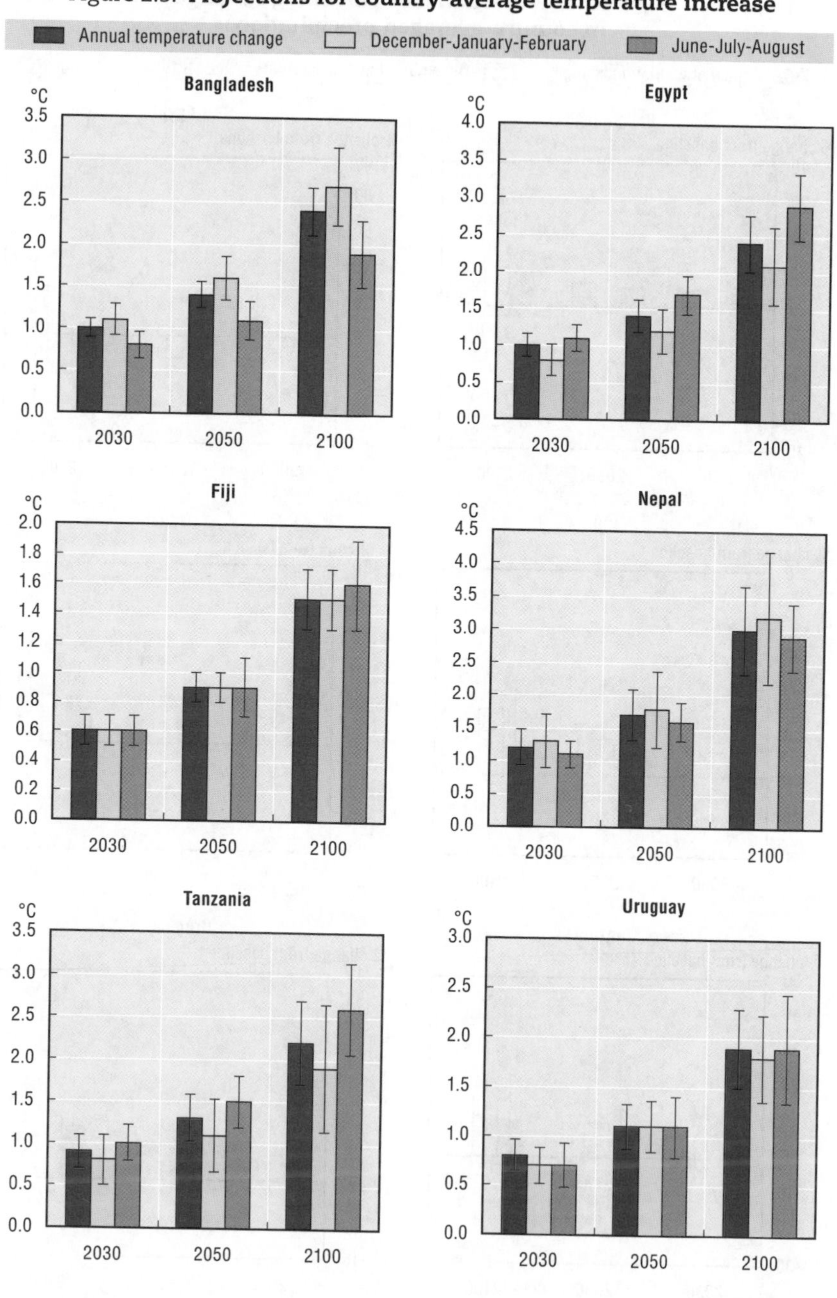

Figure 2.4. Projections for percentage change in country-averaged precipitation

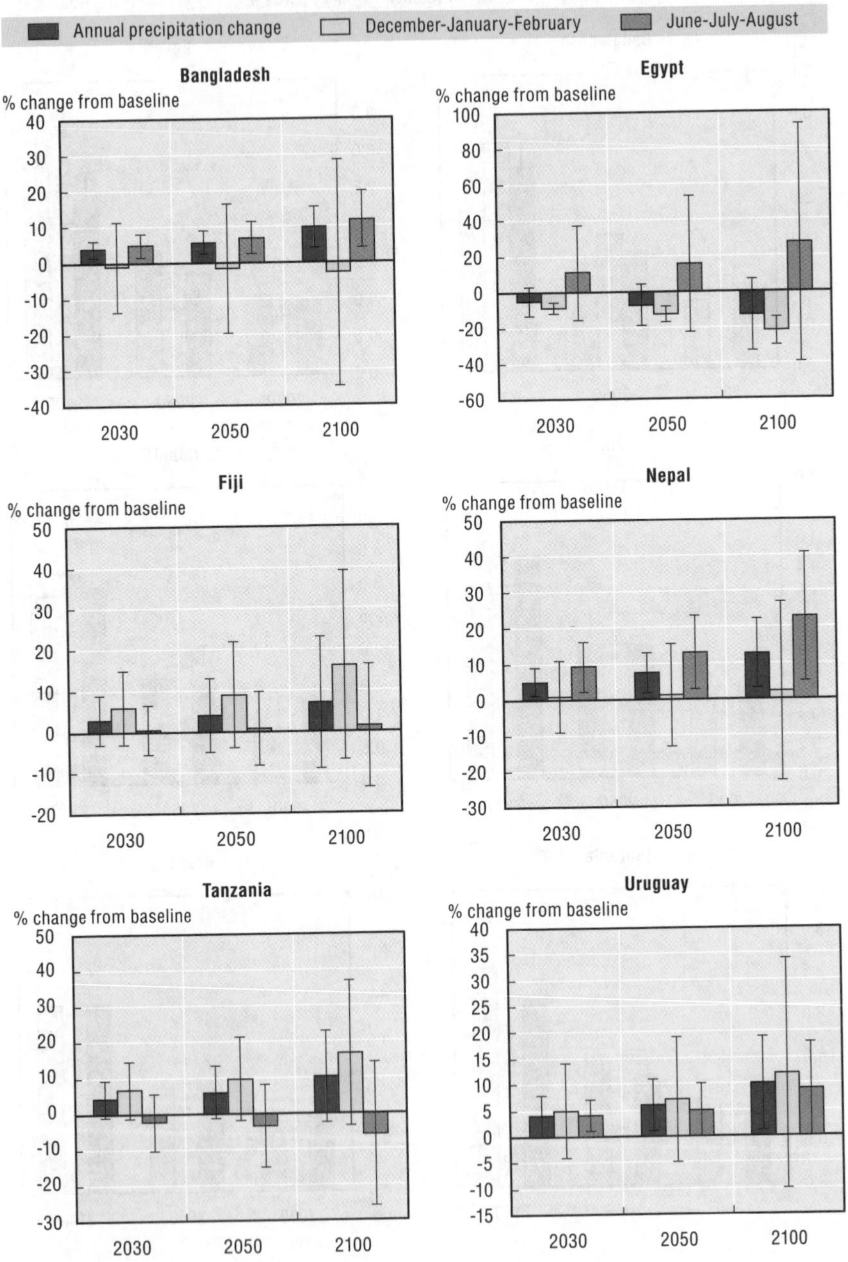

Annual precipitation change | December-January-February | June-July-August

Bangladesh

% change from baseline

40 30 20 10 0 -10 -20 -30 -40

2030 2050 2100

Egypt

% change from baseline

100 80 60 40 20 0 -20 -40 -60

2030 2050 2100

Fiji

% change from baseline

50 40 30 20 10 0 -10 -20

2030 2050 2100

Nepal

% change from baseline

50 40 30 20 10 0 -10 -20 -30

2030 2050 2100

Tanzania

% change from baseline

50 40 30 20 10 0 -10 -20 -30

2030 2050 2100

Uruguay

% change from baseline

40 35 30 25 20 15 10 5 0 -5 -10 -15

2030 2050 2100

significantly higher than in the case of impacts related more to increases in temperature, for which the agreement among models is much better.

3.2. Sea level rise

In addition to temperature and precipitation, climate change will be manifested through rising sea level, which will affect islands and coastal systems. Climate change contributes to sea level rise principally through thermal expansion of the oceans and the melting of glaciers and ice caps. The Third Assessment Report of the IPCC projects a global mean sea level rise between 9 cm and 88 cm over 1990-2100, with the central value indicating a rate of sea level rise 2.2 to 4.4 times that of the 20th century (Church *et al.*, 2001). An additional, albeit much more uncertain, possibility is that the Greenland and Antarctic ice sheets could collapse, potentially raising sea levels by as much as 6 metres. The coupled atmosphere-ocean climate model results reviewed in the IPCC Third Assessment indicate a likelihood of substantial regional variation in climate change-induced sea level rise, though confidence in projections for specific regions is low. Accordingly, regional sea level rise scenarios using climate models were not attempted in this analysis. Rather, the analysis was based upon sensitivity of coastal systems to plausible contours of sea level rise consistent with projected global mean sea levels; where possible, non-climatic factors affecting net sea level, such as local topography and vertical land movement, were also taken into account.

In Bangladesh, the Ganges-Brahmaputra delta is quite active morphologically: continuous sediment loading raises the land level even as subsidence from tectonic movements and sediment compaction has the opposite effect. A review of the literature and canvassing of expert opinion suggests that the two effects may cancel each other out, but significant uncertainty remains. Considerably compounding the impact of an increase in mean sea level is the fact that Bangladesh lies in an active cyclone corridor and is subject to frequent storm surges. Fiji, also situated in an active cyclone corridor, could expect a similar effect. While Egypt is not vulnerable to cyclones, the Nile delta's very high land subsidence rate (3-5 mm a year) will considerably exacerbate the effects of sea level rise. A cross-cutting impact of sea level rise, common to all the countries except landlocked Nepal, is the vulnerability of critical ecosystems. The coastal mangroves of Bangladesh and Fiji are particularly vulnerable, while coastal wetlands and lakes are vulnerable in Egypt, Tanzania and Uruguay.

4. Towards prioritising adaptation responses

The extent to which suitable responses to climate change are needed depends not only on the degree of certainty associated with projections, but also on the significance of resulting impacts on natural and social systems. An assessment of the vulnerability of a particular social or human system can be used to prioritise adaptation responses. Vulnerability is a subjective concept comprising three dimensions: exposure, sensitivity and adaptive capacity (Smit *et al.*, 2001). Of these, exposure is the most directly related to external stress (in this case, the biophysical impact of climate change). Sensitivity and adaptive capacity have more to do with the natural or human system at risk. No universally accepted, objective means of "measuring" vulnerability exist, but it is possible to identify vulnerable systems subjectively and assess qualitatively the extent to which vulnerability arises from a general state of deprivation or is related specifically to climate change.

Where vulnerability is primarily contextual, adaptation might require emphasis on *baseline or "business as usual"* economic development activities – alleviating poverty and improving nutrition, health care, livelihoods and so on – which will also boost the capacity for coping with climate change. Where vulnerability is significantly exacerbated by the biophysical impact of climate change, adaptation might require more explicit consideration of climate risk in development activity. For example, in the Nepal Himalayas, infrastructure and livelihoods could be catastrophically affected by glacial lake outburst floods and glacier retreat. Thus, adaptation would require more direct measures to reduce exposure to such risks, going beyond baseline poverty reduction and economic development.

In seeking to establish priorities for adaptation, decision makers may also wish to consider the affected sector's share of employment, contribution to the economy, cultural or other importance, and potential to affect other sectors. Even modest overall climate change risk affecting sectors that are central to a nation's economy or have significant distributional implications might merit higher priority than significant risk involving less critical sectors or regions.

A further possible determinant of the immediacy of adaptation responses is the period over which significant climate change impacts are expected to occur. For example, glacier retreat and permafrost melt are already having a significant impact or are projected to do so over the next few decades. Significant sea level rise, projected to occur over several decades and centuries, would in principle be less urgent. Even in this case, however, adaptation might be more urgently required in particularly low-lying areas where land is sinking because of non-climate factors, such as the Nile delta, or where storm surges could considerably amplify modest increases in sea level, as in Bangladesh.

Using several of these criteria, participants in a national workshop organised in connection with this project carried out priority setting for Nepal. Subjective scores – high, medium or low – were assigned to each sector assessed. The results were discussed and refined on the basis of insights provided by local experts, government representatives and stakeholders. Table 2.1 summarises some of the results.

BRIDGE OVER TROUBLED WATERS – ISBN 92-64-01275-3 – © OECD 2005

Table 2.1. **Priority ranking of climate change impacts for Nepal**

Sector	Certainty of impact	Timing of impact (urgency)	Severity of impact	Importance of resource
Water resources and hydropower	High	High	High	High
Agriculture	Medium-low	Medium-low	Medium	High
Human health	Low	Medium	Uncertain	High
Ecosystems/biodiversity	Low	Uncertain	Uncertain	Medium-high

Water resources and hydropower ranked significantly higher than other sectors for several reasons. First, several impacts on water resources and hydropower directly related to rising temperatures have already been observed and, according to high-confidence projections, will increase further in coming decades. They include glacier retreat, which in turn increases variability in (and eventually reduces) stream flows; and glacial lake outburst floods, which pose significant risk to hydropower facilities, other infrastructure and settlements. Such floods have already had a significant impact in Nepal, most notably the near total destruction of a new hydropower facility in 1985. Other climate-induced risks to water resources and hydropower facilities include increased intensity of precipitation, entailing flooding, landslides and sedimentation (particularly during the monsoon); and increased unreliability of dry-season flows, posing potentially serious risks of water and energy supply shortages. Further justifying the high ranking for water resources and hydropower are the significance of water resources to agriculture and the 92% share of hydropower in Nepal's electricity supply.

The impacts of climate change on other sectors tend to be less direct, less immediate, or both, and entail more uncertainty, although the sectors themselves are quite significant. Three sectors that fall in this cluster are agriculture, human health and ecosystems/biodiversity. Agriculture is a key sector in Nepal, accounting for large shares of its output and labour force. The limited information available indicates that the sector is moderately sensitive to climate change. Significant impacts may not be seen for many decades. Direct climate change risks for agriculture thus seem relatively low. Human health is ranked below agriculture mainly because of significant uncertainty about many impacts, though climate change is likely to pose health risks from increased flooding and vector-borne illness. The effects of flooding could be apparent in the near term, but other health effects might not be apparent for many decades. Ecosystems and biodiversity are ranked last because little historical research has been conducted on the effects of climate change on species diversity. It is uncertain how sensitive biodiversity will be to climate change or when the impact may be felt.

As this prototype example from Nepal illustrates, the degree of confidence in particular climate projections has significant implications for the implementation and mainstreaming of adaptation responses. The uncertainty analysis in the projections of temperature and precipitation shown in Figures 2.3 and 2.4 is a critical factor in this regard. In particular, the high uncertainty associated with precipitation projections might call into question any proposals for aggressive adaptation measures to address precipitation-related risk. While aggressive adaptation may be warranted where sufficient certainty is associated with particular impacts of climate change, other impacts might only require precautionary, "no regrets" measures to reduce vulnerability and build adaptive capacity – measures that would be warranted even if the projected changes did not occur.

BRIDGE OVER TROUBLED WATERS – ISBN 92-64-01275-3 – © OECD 2005

ANNEX 2.A1

General Circulation Model Abbreviations, Names and Citations

Model name in SCENGEN	Name	Citation	Atmospheric resolution[1]	Transient response[2] (°C)	Equilibrium climate[2] (°C)
BMRCTR98	BMRC2	Power *et al.*, 1998	R21 (3.2 × 5.6)	Not available	Not available
CCC1TR99	Canadian Model	Flato *et al.*, 2000	T32 (3.8 × 3.8)	Not available	Not available
CCSRTR96	CCSR/National Institute of Environmental Studies	Emori *et al.*, 1999	T21 (5.6 × 5.6)	1.8	3.6
CERFTR98	CERFACS	Barthelet *et al.*, 1998	T31 (3.9 × 3.9)	1.6	
CSI2TR96	Commonwealth Scientific and Industrial Research Organisation	Gordon and O'Farrell, 1997	R21 (3.2 × 5.6)	2	3.7
CSM_TR98	National Center for Atmospheric Research	Boville and Gent, 1998	T42 (2.8 × 2.8)	Not available	Not available
ECH3TR95	Max Planck	Voss, Sausen and Cubasch, 1998	T21 (5.6 × 5.6)	Not available	Not available
ECH4TR98	ECHAM4 + OPYC3	Roeckner *et al.*, 1996	T42 (2.8 × 2.8)		
GFDLTR90	Geophysical Fluid Dynamics Laboratory	Manabe *et al.*, 1991	R15 (4.5 × 7.5)	Not available	3.7
GISSTR95	Goddard Institute for Space Studies	Russell, Miller and Rind, 1995	(4.0 × 5.0)	1.5	Not available
HAD2TR95	United Kingdom Meteorological Office	Johns *et al.*, 1997	(2.5 × 3.75)	Not available	Not available
HAD3TR00	United Kingdom Meteorological Office	Gordon *et al.*, 2000	(2.5 × 3.75)	Not available	Not available
IAP_TR97	IAP/LASG2	Zhang *et al.*, 2000	R15 (4.5 × 7.5)	Not available	Not available
LMD_TR98	LMD/IPSL2	Leclainche *et al.*, in review	Not available	Not available	Not available
MRI_TR96	MRI2	Tokioka *et al.*, 1996	(4.0 × 5.0)	1.6	2.5
PCM_TR00	National Center for Atmospheric Research Parallel Climate Model	Washington *et al.*, 2000	T42 (2.8 × 2.8)	1.3	2.1
W&M_TR95	National Center for Atmospheric Research	Washington and Meehl, 1996	R15 (4.5 × 7.5)	Not available	Not available

1. Grid cells in degrees. Result reported as available from McAveney *et al.* (2001).
2. From Cubasch *et al.* (2001).

BRIDGE OVER TROUBLED WATERS – ISBN 92-64-01275-3 – © OECD 2005

ISBN 92-64-01275-3
Bridge Over Troubled Waters
Linking Climate Change and Development

Chapter 3

Analysis of Donor-supported Activities and National Plans

by

Maarten van Aalst and Shardul Agrawala

Climate change can affect the efficiency with which development resources are invested and the eventual achievement of many development objectives. Hence the need for mainstreaming of climate change response measures in development initiatives. This chapter examines government- and donor-supported initiatives in each case study country. An analytical framework is developed to quantify exposure of development aid portfolios to climate risk using the Creditor Reporting System database, which provides standardised information on aid flows. This is followed by an analysis of high-level strategy documents, sectoral plans and project documents to assess the degree of attention being paid to climate change impacts and adaptation.

1. Introduction

Decisions made as part of development activities can have considerable bearing on future GHG emissions as well as the vulnerability of societies to the potential impact of climate change. The latter issue is the principal focus of this chapter. In principle, a range of activities oriented towards reducing poverty, improving nutrition and education and promoting sustainable livelihood opportunities would help reduce vulnerability to many climate change impacts. A healthier, better-educated population with improved access to resources is likely to be in a better position to cope with climate change.

In many cases, however, climate change can affect the efficiency with which development resources are invested and the eventual achievement of many development objectives. Hence the need to explicitly incorporate or "mainstream" climate change considerations within a range of development activities. This is particularly so for policies and projects with implications over the longer term, when many climate change impacts will manifest themselves. As activities by governments and development co-operation agencies often have long time horizons, they are especially relevant. Private investment, particularly foreign direct investment, has also become important for developing countries, particularly those in the upper middle income category (Figure 3.1 and Box 3.1, which explains terms and abbreviations used in this discussion). For the least developed countries as well as those in the low and middle income categories, official flows – grants and loans – are much more significant, and thus a higher priority area for mainstreaming.

This chapter examines government- and donor-supported initiatives in the six developing countries for which case studies were carried out: Bangladesh, Egypt, Fiji, Nepal, Tanzania and Uruguay. In terms of scope, a primary focus is on attention given to climate impacts and the mainstreaming of climate adaptation in development activities, although there is some consideration of links between development priorities and resulting implications for GHG mitigation.

Section 2 describes the analytical framework used to assess exposure of development aid portfolios to climate risk and examines the degree of attention to climate change in donor country assistance strategies and projects. Exposure of donor-supported activities is quantified using the Creditor Reporting System (CRS) database, which provides standardised information on aid flows by donor, activity and country. Activities undertaken by national governments are also potentially affected by climate change. Therefore, mainstreaming

BRIDGE OVER TROUBLED WATERS – ISBN 92-64-01275-3 – © OECD 2005

Figure 3.1. **Official and private financial flows to developing countries**

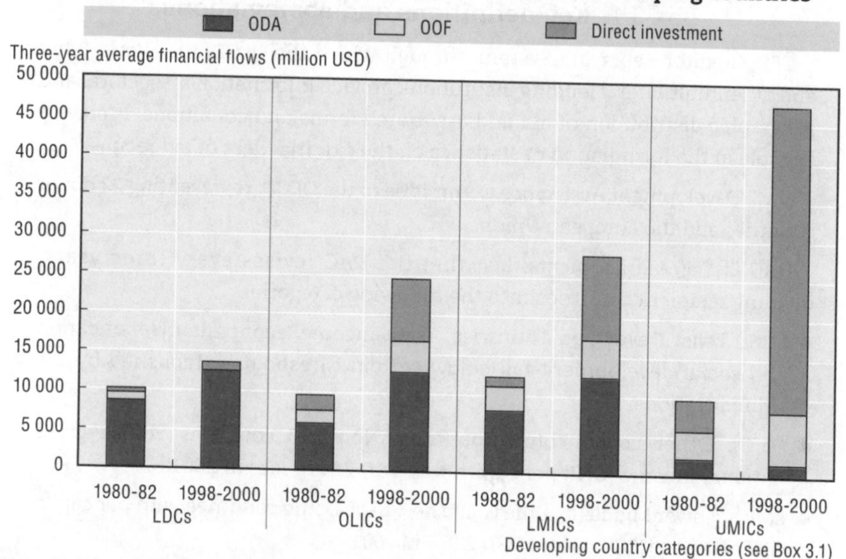

Note: Based on DAC database Table 4: Destination of Private Direct Investment and other Private Capital (aggregates only); Table 2a: Destination of Official Development Assistance and Official Aid – Disbursements; Table 2b: Destination of Other Official Flows – Disbursement (aggregates only). OECD (2004c), last accessed in April 2004.

within national development plans and projects may also be needed. Section 3 assesses the attention to climate change in government- and donor-supported activities. Section 4 provides some concluding remarks.

2. Analysis of development aid statistics

Donor agencies are increasingly interested in incorporating climate change concerns in their core development activities. The World Bank, GTZ and NORAD, to cite some prominent examples, have reported on links between their development assistance activities and climate change adaptation (Burton and van Aalst, 1999 and 2004; Klein, 2001; Eriksen and Næss, 2003). The analysis in this chapter is the first attempt to review development strategies and projects for several bilateral and multilateral development agencies.

A comprehensive evaluation of the extent to which development activities are affected by climate change would require detailed assessments of all relevant projects and consideration of site-specific climate change impacts, both of which are beyond the scope of this analysis.[1] Instead, this chapter seeks to determine what proportion of total aid portfolios may be in sectors potentially affected by climate risk, where climate change concerns thus may need to be taken into account. This is accomplished through an analysis of CRS data on ODA and other official aid.

Box 3.1. **Key definitions and abbreviations**

CRS: Creditor Reporting System, through which the OECD, the World Bank and other multilateral lending institutions provide information on sectoral and geographic distribution of aid and associated terms and conditions. It plays a key role in the formulation of statistics on the external debt of aid recipients.

DAC: Development Assistance Committee of the OECD, representing 22 donor countries and the European Union.

DAC List of Aid Recipients: List that the DAC revises every three years, dividing recipient countries into the following categories:

- *LDCs:* Least Developed Countries. Their income, economic diversification and social development fall below certain thresholds established by the United Nations.
- *OLICs:* Other Low Income Countries. Non-LDC countries whose gross national income (GNI) per capita was USD 745 or less in 2001.
- *LMICs:* Lower Middle Income Countries. Developing countries with per capita GNI between USD 746 and USD 2 975 in 2001.
- *UMICs:* Upper Middle Income Countries. Developing countries with per capita GNI between USD 2 976 and USD 9 205 in 2001.
- *HICs:* High Income Countries, with per capita GNI exceeding USD 9 206 in 2001. As of 2004 the only developing country in this category was Bahrain.

FDI: Foreign Direct Investment. Private sector investment made to acquire or add to a lasting interest in an enterprise in a country other than that of the investor. "Lasting interest" implies a long-term relationship in which the investor has a significant influence on the management of the enterprise, reflected by ownership of at least 10% of the shares, equivalent voting power or other means of control.

Grant: Transfer made in cash, goods or services for which no legal debt is incurred by the recipient.

Official flows: All transactions from the official sector to developing countries.

ODA: Official Development Assistance. Grants and concessional loans to developing countries by the official sector whose main objective is the promotion of economic development and welfare. Loans must include a grant element of at least 25% to qualify as ODA.

OOF: Other Official Flows. Transactions by the official sector with countries on the DAC List of Aid Recipients which do not meet the conditions for eligibility as ODA or official aid, either because they are not primarily aimed at development, or because they have a grant element of less than 25%.

Source: OECD (2003 and 2003e).

BRIDGE OVER TROUBLED WATERS – ISBN 92-64-01275-3 – © OECD 2005

Aid flows vary from year to year. To even out some of the effects of this variation, the analysis uses a recent three-year sample of aid commitments from the CRS database, for 1998-2000. To account for all official aid flows, the analysis includes both ODA and OOF. Table 3.1 provides an overview of the components of official flows for the six case study countries. Official aid is clearly a significant factor in development activities, with Bangladesh and Egypt receiving over USD 1.5 billion annually. While Fiji receives a more limited amount, averaging USD 28.5 million, this aid is still quite significant relative to the size of its economy and population. For most of the case study countries, the ODA tends to dominate. The exception is Uruguay, which receives little ODA but a significant amount of OOF.

Table 3.1. **Overview of annual official flows to case study countries, 1998-2000**

Million USD

	Total official flows	ODA	OOF
Bangladesh	1 766.0	1 739.0	27.0
Egypt	1 608.0	1 471 .0	137.0
Fiji	28.5	28.0	0.5
Nepal	320.0	320.0	0
Tanzania	972.0	972.0	0
Uruguay	336.0	9.0	327.0
Total	**5 030.5**	**4 539.0**	**491.5**

Source: OECD (2004c).

The assessment of the proportion of aid flows that might be affected by climate change is accomplished through an analysis of aid commitments to specific activity areas. The CRS database maps all development assistance activities in terms of "purpose codes" identifying the area of the recipient country's economic or social structure at which the transfer is aimed. A subset of these codes includes activities that could be affected by climate risk and might therefore need to take climate concerns into account. They could range from initiatives to promote agriculture in areas that might become more, or less, viable under climate change, to infrastructure investments that could be at risk from impacts such as permafrost melt, glacier retreat and sea level rise. Also included are projects that affect the vulnerability of other natural or human systems to climate change. For instance, new roads might be fully weatherproof from an engineering standpoint, even taking future climate into account, but they might trigger new human settlement in areas at high risk for particular impacts of climate change, such as coastal zones vulnerable to sea level rise. Such considerations might also need to be taken into account in project design and implementation. By contrast, development activities related to education, gender equality and governance would be much less directly affected by climate change.

Any classification of this type will be oversimplified. The reality is a wide spectrum of exposure to climate risk even within a given sector or purpose code area. For instance, agriculture projects in rain-fed areas might be much more vulnerable than projects in areas with reliable irrigation. While most education projects would not be affected by climate risk, school buildings in flood-prone areas might well be vulnerable. Without an in-depth examination of risks to individual projects, it is impossible to capture such nuances. The present classification provides a rough sense of the share of development activities that might be affected by climate risk. If an activity falls within the "climate-affected" cluster, that does not mean it needs to be redesigned in the light of climate change, or that it is possible to quantify current and future climate risk. The only implication is that climate risk could well be one of many factors to take into account in activity design – a marginal factor in some cases, a substantial one in others. Either way, the activity would benefit from an initial screening for the significance of climate risks. Hence, one might expect to see attention paid to them in project documents and in sector strategies or national development plans.

To incorporate some of the uncertainty inherent in the sectoral classification, the share of development activities potentially affected by climate change was calculated in two ways, one rather broad and the other more restrictive. The first selection includes projects dealing with infectious diseases, water supply and sanitation, transport and storage, agriculture, forestry and fisheries, renewable energy and hydropower, tourism, urban and rural development, environmental protection, food security and emergency assistance. The second selection excludes projects related to transport and storage, as well as food aid and emergency assistance. Transport and storage projects make up a relatively large share of the development portfolio in many countries because the individual investments required are large, unlike in "softer" sectors such as environment, education and health. At the same time, however, infrastructure projects are usually designed on the basis of detailed engineering studies, which should already include attention at least to current climate risk. As for food aid and emergency assistance, such activities are generally reactive and planned at short notice. The treatment of risk is thus very different from well-planned projects intended to have long-term development benefits.

Together, the first and second selections give an indication of the share of climate-affected development activities. Annex 3.A1 shows the CRS purpose codes for activities included in the category of climate-affected projects; activities excluded from the more restrictive second selection are marked with a superscript.

Figure 3.2 presents the results of the analysis. The height of each bar indicates the total of annual official flows to the country, averaged over 1998-2000, and the dark portion indicates flows directed to activities potentially affected by climate change. The shading reflects the range of estimates captured in the broad and narrow selections.

Figure 3.2. **Annual official flows and share of activities potentially affected by climate change**

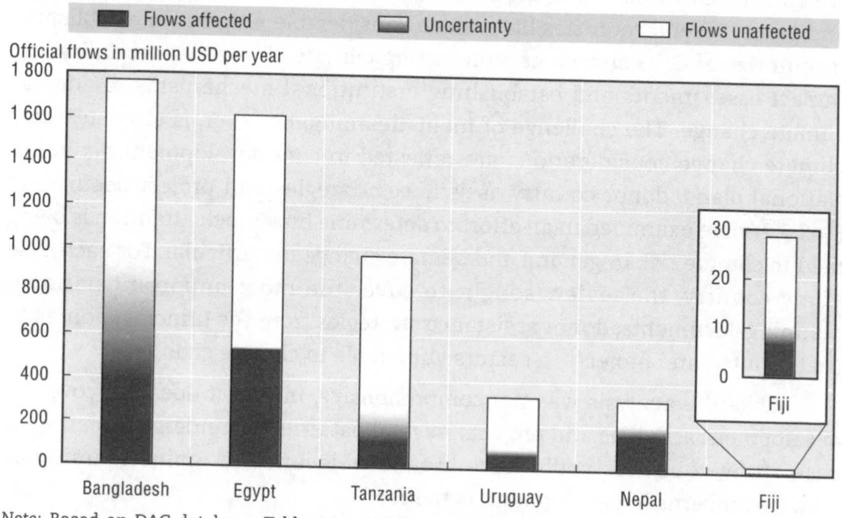

Note: Based on DAC database, Table 4: Destination of Private Direct Investment and other Private Capital (aggregates only). OECD (2004c), last accessed in March 2004.

The analysis implies that a large share of development assistance might be in activities affected by climate risk: from as high as over half a billion dollars annually in Bangladesh and Egypt to around USD 200 million in Tanzania and Nepal. In Fiji, while the absolute amount may be low, it constitutes roughly one-third of all aid flows. Uruguay is the exception because it receives very little ODA, as a UMIC: most of its official flows are loans, primarily in activities not directly exposed to climate risk.

It is evident that consideration of climate risk (including climate change) could be central to the achievement of general development goals as well as the success of individual investments and projects. The amount of official aid flows in activities potentially affected by climate risks is considerably higher than funding committed to financing climate change adaptation *per se*. Thus, over the long term, there appears to be far greater potential for mainstreaming adaptation within core development activities than for financing of action on adaptation within the climate regime. Another implication of the analysis relates to the question of the optimal use of limited climate change adaptation funds: one possibility might be to use them to trigger mainstreaming of adaptation, particularly in the context of development activities by the GEF implementing agencies.

3. Analysis of development plans and projects

The potential significance of climate change for core development activities, as illustrated above, highlights the need to assess the extent to which such concerns are being addressed in current development strategies and plans. Several countries have made considerable progress in establishing inventories of GHG emissions, conducting climate change vulnerability and impact assessments and establishing institutional mechanisms to address climate change. The challenge of mainstreaming, however, is to ensure that climate change considerations are reflected in core development activities. National plans, donor country assistance strategies and project documents (Box 3.2) were examined in an effort to determine how much attention is being paid to climate risk in general and climate change in particular. For each case study country, the review sought to cover the most important national planning documents, donor assistance strategies from the principal donors to that country, and projects in sectors vulnerable to climate change.

While the analysis was not comprehensive, in that it does not cover all development activities and projects, several patterns that emerge from the six case study countries could have broader relevance to mainstreaming of climate concerns in development activities.

Box 3.2. **Development plans and projects examined**

National documents

- National development plans, Poverty Reduction Strategy Papers (PRSPs).
- Sectoral development strategies, National Strategies for Sustainable Development.
- National communications under the UNFCCC, the Convention on Biological Diversity (CBD) and the United Nations Convention to Combat Desertification (UNCCD).

Donor strategies

- Country assistance strategies, country strategy papers, sectoral development strategies.

Project documents

- Project design documents, project evaluations, environmental impact assessments.

BRIDGE OVER TROUBLED WATERS – ISBN 92-64-01275-3 – © OECD 2005

3.1. National policies and plans specific to climate change have made considerable inroads

The case study countries are all very active in international efforts to address climate change. All six have ratified the UNFCCC. All but Nepal have also signed the Kyoto Protocol. As of mid-2004, Fiji and Uruguay had ratified the Kyoto Protocol, and Bangladesh and Tanzania had deposited instruments of accession. Bangladesh, Egypt, Tanzania and Uruguay had deposited their first national communications with the UNFCCC Secretariat, and those of Nepal and Fiji were unavailable at the time of writing this chapter but expected shortly.

Countries have established varying numbers of domestic plans and institutional mechanisms to co-ordinate and, in some cases, to implement activities on climate change. For example, *Egypt* launched two major programmes in 1995 to respond to the UNFCCC: Support for National Action Plan and Building Capacity for Egypt. The programmes have facilitated 25 studies covering GHG emission inventories, GHG mitigation technology assessment, abatement cost analysis, and impacts and adaptation assessment. In addition, Egypt established an inter-ministerial National Climate Change Action Committee in 1997. *Tanzania*'s 1997 National Action Plan on Climate Change outlines a three tier strategy, focusing in the short term (one to two years) on awareness raising; in the medium term (two to five years) on support for projects internalising climate change aspects, particularly GHG mitigation, inclusion of climate change issues in the educational curriculum and introduction of economic instruments to address climate change; and in the long term (up to 20 years) on large projects in the energy and transport sectors as well as structural adaptation to cope with sea level rise.

All case study countries have made considerable progress with regard to attention to climate change. This progress is perhaps strongest in terms of climate change assessments and the establishment of institutional mechanisms to conduct such work. Both have been spurred in large part by the commitment to produce national communications under the UNFCCC. The institutional mechanisms are generally handled by environment ministries, with limited involvement from other ministries. Analysis related to mitigation scenarios and measures tends to be much more detailed and sophisticated than that on impacts and adaptation.

Overall, progress in terms of actual actions to mitigate or adapt to climate change has been very limited. For example, *Tanzania* has made no discernible progress on implementing the medium-term objectives outlined in its 1997 action plan. That plan also outlined adaptation responses for water resources, with a particular focus on water conservation. Similar objectives with regard to water conservation had been set out in the 1991 National Water Policy, but remain unmet. Clearly, implementation remains a principal bottleneck, even where specific climate change action or related sectoral plans exist (OECD, 2003d).

3.2. *Donors are actively involved in many climate change plans and projects*

Like their national government counterparts, development co-operation agencies are involved in climate-related activities in the case study countries. International donors have played a significant role in the preparation of national communications through bilateral country study programmes, such as those of the United States and the Netherlands. Donors also channel resources through the GEF to support activities that build capacity for developing national communications. Some donors are working with national governments and other partners in projects and programmes in areas ranging from energy efficiency and GHG mitigation to adaptation and coastal zone management. The individual case study reports (OECD, 2003a-d and 2004a-b) discuss these activities.

A number of initiatives are also under way to better integrate climate change concerns into development co-operation activities. Climate specialists and programme managers from ten development agencies (led by the World Bank) wrote *Poverty and Climate Change: Reducing the Vulnerability of the Poor through Adaptation*, a report highlighting the need to integrate climate change concerns in poverty alleviation efforts (Sperling, 2003). The European Commission has developed a climate change strategy for support to developing countries (European Commission, 2003) whose overall objective is to mainstream climate concerns into EU development co-operation so as to help countries meet challenges posed by climate change. Its strategic priorities are: *i)* raising the policy profile of climate change; *ii)* providing support for adaptation; *iii)* providing support for mitigation; and *iv)* developing capacity. The inter-agency report and the EU strategy outline general objectives not specific to particular country contexts. They were too recent at the time of writing for their operational impact on in-country development co-operation policies to be assessed.

Among the case study countries, the most significant development co-operation efforts on climate change adaptation, and on linking such responses with development policies, have been in Bangladesh and Fiji.

In *Bangladesh*, the World Bank report *Bangladesh: Climate Change and Sustainable Development* reviewed 16 development co-operation activities in the light of adaptation to climate change. The activities were carried out mainly by the World Bank and the Asian Development Bank (ADB), but also by the Netherlands and the United Kingdom. The report's main finding was that development activities did not generally consider climate change impacts or adaptation. It highlighted mainstreaming of climate change concerns as a priority for future national planning and development assistance activities (World Bank, 2000a). Recent reviews by Huq (2002) and Rahman and Alam

(2003) indicate that this report has resulted in some effort at mainstreaming, particularly in a World Bank coastal management project and the GEF/ADB biodiversity conservation project in the Sundarbans.

In *Fiji*, several programmes have focused on sea level rise monitoring and impacts assessment. Examples are AusAid's climate change and sea level rise monitoring programme and a coastal zone management project by the Environment Agency of Japan and the Pacific Regional Environment Programme (SPREP). The GEF-funded Pacific Islands Climate Change Assistance Programme (PICCAP) was also quite successful in mapping climate change impacts and identifying adaptation options. While PICCAP's early focus was on monitoring, it later moved towards implementation of risk reduction policies and measures. AusAid's monitoring programme is moving into its third phase, which includes planning of response and adaptation measures. AusAid has established a regional adaptation fund that can finance pilot projects.

A more direct move towards adaptation and mainstreaming in Fiji came with the World Bank assessment *Cities, Seas, and Storms: Managing Pacific Island Economies* (2000). As part of this assessment, the World Bank produced a volume on *Adapting to Climate Change*, with a particular focus on Fiji and Kiribati. Like the Bangladesh report, it placed responses to climate change within the context of development planning, noting that "the development choices made by Pacific Island governments today will have a profound impact on the future vulnerability of the islands and the magnitude of climate change impacts" (World Bank, 2000b). The report established guidelines for selecting adaptation measures and highlighted a two-track approach to implementation that focused on mainstreaming and building partnerships.

While the World Bank has not initiated projects on adaptation and mainstreaming in Fiji, it has made considerable progress in Kiribati, working closely with several government ministries, other donors and a range of other stakeholders. An aim of the project, which has included national adaptation consultations and capacity-building activities, is to incorporate a national adaptation "vision" into the 2004-08 national development strategy (van Aalst and Bettencourt, 2004). A key feature of this effort is the involvement of "line ministries": the Ministry of Finance and Economic Planning, for instance, is the executing agency for the project. While implementation is only just getting under way, the Kiribati project may represent the cutting edge in terms of donor-government collaboration on mainstreaming responses to climate change. Meanwhile, the Canadian International Development Agency recently initiated an effort with an explicit mainstreaming objective in Fiji and three other Pacific Island countries. It is too early to assess the project or the extent to which it will involve key stakeholders and government agencies concerned with economic development and sectoral planning.

3.3. *National reports on other environmental concerns pay limited attention to climate change*

The real challenge for mainstreaming is to increase integration of climate change concerns in development activities that are not specific to climate but could be affected by climate change. An initial area of focus is related transboundary and global environmental issues that could intersect with climate change, such as wetland protection, biodiversity conservation and efforts to combat desertification. Wetland protection measures, for example, might need to take into account the additional risk posed by sea level rise and saline intrusion. Biodiversity conservation could be affected by species migration and other impacts of climate change on forests and biodiversity.

The countries examined in this analysis have produced reports under environmental agreements such as the Ramsar Convention on Wetlands, the CBD and the UNCCD. Many of these reports mention climate change, though some overlook it. For example:

- *Fiji*'s report under the CBD discusses threats to coastal ecosystems, including salinity changes and sedimentation due to flooding and cyclones, as well as the negative impacts of coastal development, including seawalls, land reclamation, dredging and sedimentation due to clearing of land for agriculture. Potential climate change impacts, particularly sea level rise, are not discussed.
- *Tanzania*'s report under the CBD does not mention climate change, but its second national report under the UNCCD does highlight links between climate change and desertification. That report also notes that programmes on desertification have been quite successful in terms of awareness raising and in mainstreaming desertification concerns in national and sectoral plans and policies.
- *Uruguay*'s second report under the UNCCD raises concerns about climate change, particularly given the impacts of current weather and climate events (including the El Niño and La Niña phenomena). It also draws attention to GHG mitigation activities and new opportunities to utilise the Clean Development Mechanism.
- *Nepal*'s 2002 Biodiversity Strategy lists several climate-related risks, including flooding and siltation, as threats to biodiversity conservation. The strategy does not, however, discuss the possibility that some risks, including both flooding and siltation, could be exacerbated significantly under climate change.

Documents prepared by the case study countries for the 2002 World Summit on Sustainable Development in Johannesburg either treat climate change in isolation or omit it:

- *Bangladesh*'s documentation discusses climate change as a stand-alone air quality issue rather than a cross-cutting concern affecting many aspects of sustainable development.
- *Egypt*'s country profile reflects an isolation of climate issues from mainstream development planning: climate change comes up only in a separate section on protection of the atmosphere, and then only in terms of reducing GHG concentrations.
- *Tanzania* refers to its vulnerability and adaptation assessment, listing agriculture, water resources, forestry, grasslands, livestock, coastal resources, and wildlife and biodiversity as vulnerable to climate change. It notes that efforts to address current vulnerability to climate-related risks include several components of potential climate change adaptation strategies, such as better water management (in the context of irrigation development) and research on drought-resistant, high-yield crops. Tanzania devotes little attention to adaptation (except in agriculture, where further work is planned) but discusses mitigation extensively.
- *Uruguay*'s report outlines a strategy aimed at providing a healthy environment capable of productively sustaining quality of life. The strategy includes scientific and technological proposals, proposals for managing natural risk and reducing vulnerability, and a proposal for the recovery and management of coastal areas.
- *Nepal*'s country profile discusses climate change only in the context of GHG emission mitigation; adaptation is not mentioned. However, the section on sustainable mountain development does pay attention to indigenous systems of human adaptation to geographic and climatic challenges. Many elements of proposed sustainable development policies (designed for current climate circumstances) would also be "no-regrets" adaptation measures.

3.4. Core national development plans do not recognise climate change

Perhaps even more critical than mainstreaming climate change in related environmental issues is the extent to which climate change concerns are reflected in core development activities. Most of the case study countries have general development plans, often five-year plans but sometimes with a longer horizon. While it is natural for such documents to focus on immediate social and economic priorities, current climate risks and long-term climate change have implications for the achievement of objectives. In general, even long-term national planning documents do not mention climate change. Current climate risks are occasionally mentioned, but generally with no explicit consideration of how to account for them in meeting development objectives.

- *Tanzania*'s 1998 long-term strategy, National Development Vision 2025, highlights objectives including economic prosperity, equity, self-reliance, economic diversification and industrialisation, and sustainability. The document does not mention climate change, though it discusses climate-related risks, mainly floods and droughts.
- In 2002, *Nepal*'s National Planning Commission adopted its tenth development plan (2002-07), with poverty alleviation as its primary objective. It includes programmes in agriculture, tourism, financial services and industry, electricity and fuels, social services, rural infrastructure and governance, with an overall goal of reducing the share of the population in poverty to below 30%. While many of the development activities proposed could well reduce vulnerability to climate risks, explicit attention to these risks is lacking. The plan does not explore ways to reduce climate risks or analyse the risks themselves.
- *Fiji*'s Strategic Development Plan does not pay explicit attention to climate change, but it addresses many environmental vulnerabilities related to current and future climate risks. It recognises that while Fiji has a "generally benign climate" it can experience "climatic extremes in the form of hurricanes, cyclones, floods, and drought", with "serious economic, social, and environmental consequences that require prudent macroeconomic management, proper land use planning, and water and watershed management". The plan lists natural disasters among the key risks to the economy. It notes that environmental vulnerability is caused not just by natural factors but also by ineffective handling of problems such as land degradation, climate change, increasing flood risk, unsustainable exploitation of marine resources, waste management, air and water pollution, and environmental consequences of urbanisation. Concerning disaster risk reduction, the plan advocates risk assessment for urban centres, a database of infrastructure disaster mitigation priorities, reduction of land degradation and fires, and promotion of traditional cropping systems to enhance the resilience of small communities.

3.5. Donor country assistance and sectoral strategies generally do not recognise climate change

The limited degree of explicit attention to climate change in national plans is reflected in high-level policy documents of the principal donors to the six case study countries. The main differences among country assistance strategies and other such documents are reflected in their degree of attention to current climate risks.

BRIDGE OVER TROUBLED WATERS – ISBN 92-64-01275-3 – © OECD 2005

The World Bank's 2020 Long-run Perspective Study on *Bangladesh*, written in the mid-1990s, is among the very few mainstream donor documents to raise the issue of climate change. It states that "although the impacts of global warming are still far from precisely predictable, the prospect is sufficiently likely and alarming to warrant precautionary action at the national level and at the international level". The report advocated a dual response: international efforts towards global mitigation and national planning for adaptation. This was followed in 2000 by *Climate Change and Sustainable Development*, discussed above. The World Bank's country assistance strategy, published a year later, makes little mention of climate change, however, though it pays ample attention to current natural hazards.

A similar pattern occurs in high-level policy documents of the principal multilateral and bilateral donors for other case study countries, such as *Egypt*, *Tanzania* and *Nepal*. There is frequent recognition of the impact of current weather risks on development prospects, as well as mention of pressing societal problems such as vulnerability of the agricultural sector and water availability, but no discussion of the implications of climate change on such problems. However, the World Bank country assistance strategy for *Egypt* does emphasise some policy priorities that are consistent with adaptation to potential impacts of climate change on water availability. These priorities include highlighting the need for co-operation with other Nile Basin countries and reducing water consumption by improving the operation and maintenance of drainage and irrigation systems and by rationalising water allocation. The strategy also suggests a move out of water-intensive crops like rice and sugarcane.

In some cases, donor documents do not mention current climate risks, even in contexts such as discussion of development issues in mountain areas in *Nepal* where climate-induced glacier retreat and glacial lake outburst flooding already pose significant risks to society and to development infrastructure.

3.6. PRSPs do not recognize climate change but often mention vulnerability to climate risks

The World Bank and the International Monetary Fund launched the PRSP initiative in 1999. These strategy papers are produced by heavily indebted poor countries (or other countries seeking concessional loans) with World Bank/IMF assistance and participation by civil society. PRSPs, which describe macroeconomic, structural and social policies to reduce poverty and promote growth, have gained broad acceptance among the principal donors and are seen as providing the basis for most future multilateral and bilateral lending. Given their critical importance to national development planning and donor lending priorities, it is useful to examine the extent to which they consider climate risks and climate change concerns. Of the six countries under consideration here, the PRSP process is furthest along in *Tanzania*, which produced an Interim-PRSP

(I-PRSP) in early 2000 and a PRSP later that year, and has since prepared two progress reports. *Nepal* completed its PRSP in November 2003. *Bangladesh* completed an I-PRSP in June 2003. *Egypt*, *Fiji* and *Uruguay* are not preparing PRSPs. Climate change considerations and how they might affect the achievement of poverty alleviation objectives do not appear to be an explicit priority in the drafting or review of PRSPs.

Tanzania's PRSP recognises that weather and climate hazards have a serious impact on development and on the poor. It does not mention potential risks posed by climate change. The PRSP lists activities to reduce vulnerability to current climate risks, including early warning systems, irrigation, improvements to food supply systems, development of drought-resistant crops, an increase in adequate water supply from 48.5% of the rural population in 2000 to 85% by 2010, promotion of rainwater harvesting, sustained reforestation efforts and sustained efforts in adaptation. The first PRSP progress report noted that agricultural growth lagged behind expectations and cited "adverse weather and the collapse of export prices". The proposed responses to this delay do not include direct measures to reduce vulnerability to climate risks. Nor do the World Bank/IMF joint staff assessments of the PRSP and the progress report discuss such considerations.

Nepal uses its tenth development plan as its PRSP. The document does not examine climate-related risks to poverty reduction or development, or risks posed by climate change. These omissions are particularly significant because Nepal has already experienced significant temperature increases in the high Himalayas, glacier retreat and expansion of glacial lakes, which have significant implications for development. The World Bank/IMF assessment of Nepal's PRSP in 2003 does not discuss current or future climatic risks either but focuses on risks resulting from internal conflict, institutional weaknesses and failure of public service delivery.

Bangladesh's I-PRSP recognises the direct links between poverty and vulnerability to natural hazards, which are likely to increase under global climate change. It proposes a comprehensive and anticipatory approach to reducing this vulnerability. Apart from a discussion on disaster trends, however, climate change is given limited consideration in the context of planning vulnerability reduction measures.

3.7. Some sectoral policies have synergies with climate change responses; a few mention climate change explicitly

While national development and donor strategies do not generally pay explicit attention to climate change, responses to climate change are often synergistic with measures that might be undertaken in response to other sectoral priorities. In a few instances, sectoral policies also explicitly take climate change considerations into account.

BRIDGE OVER TROUBLED WATERS – ISBN 92-64-01275-3 – © OECD 2005

In *Fiji*, several sectoral plans contain examples of appropriate adaptation strategies, including (in agriculture) promotion of non-sugar crops and commodities that will enhance food security, (in forestry) a switch to sustainable management strategies and (in fisheries) a moratorium on reef mining and a review of the Mangrove Management Plan, since mangrove depletion is already hurting coastal fisheries.

Tanzania has a number of environmental and sectoral policies and plans, largely put in place during the 1990s, that are intended to increase its ability to cope with current environmental problems and the additional risks posed by climate change. For example, the National Environmental Policy of 1997 provides a framework for mainstreaming environmental considerations into decision-making processes. Concerning forestry, the policy concludes: "the main objective is the development of sustainable regimes for soil conservation and forest protection, taking into account the close linkages between desertification, deforestation, freshwater availability, climate change, and biological diversity."

An analysis of *Nepal*'s sectoral Medium Term Expenditure Frameworks (MTEF) gives the impression that climate risks in general tend to be neglected in the country's development policy. For example, the MTEF paper for the power sector does not recognise risks to hydropower plants from run-off variability, flooding (including glacial lake outburst flooding) and sedimentation. The MTEF paper for the agriculture sector does pays some attention to climate-related risks. For instance, it mentions that the monsoon season is critical for the sector.

Bangladesh's national water policy (1999) and national water management plan (2001) both recognise climate change as a factor in Bangladesh's future water supply and demand. Some implementation priorities aimed at coping with alternating flooding and droughts have synergies with climate change adaptation measures. Examples of the former include development of an early warning and flood-proofing system; comprehensive development and management of the main rivers through a system of barrages; and enhancement of regional co-operation among co-riparian countries. In addition, some elements of the national forest policy, such as development of coastal green belts, would be good "no-regrets" adaptation responses to reduce coastal vulnerability to cyclones and storm surges (Ahmed, 2003; OECD, 2003b).

Uruguay has adopted policies and legislation for agriculture and forestry that have considerable synergy with GHG mitigation objectives. The 1982 Soil Management Law fostered widespread use of soil conservation techniques and resulted in average annual sequestration of 1.8 million tonnes of carbon. Another key innovation was the Forestry Promotion Policy of 1987, which provided financial incentives to encourage plantation on soil with low agricultural productivity and/or high susceptibility to erosion. The legislation underlying the policy (Law No. 15939) explicitly recognises the "climate benefits" of forests – a particularly significant point since the law was passed before climate change had become an international policy concern. The impact of the policy has been remarkable: the forest plantation area increased from about 200 km^2 in 1987 to over 6 500 km^2 in 2000. The cumulative net carbon sequestration from 1988 to 2000 is estimated at 27.4 million tonnes of CO_2 (Baethgen and Martino, 2004; OECD, 2004b).

Some potential for conflict between sectoral development policies and climate change considerations also exists. The conflict between policies for GHG mitigation and for accelerated development based on fossil fuel consumption are well known. Some sectoral policies can promote maladaptation or exacerbate vulnerability to climate change impacts. For example, in *Bangladesh*, policies to encourage tourism and build tourism infrastructure in vulnerable coastal areas, particularly in the Khulna region, may need to take into account the projected impact of sea level rise and coastal flooding to reduce the risk of maladaptation. Similarly, plans to encourage ecotourism in the fragile Sundarbans may pose a risk of added stress to a fragile ecosystem that is projected to be critically affected by sea level rise and salinity. An analogous problem exists in *Tanzania*, whose Kilimanjaro ecosystem is vulnerable to forest fires as a result of warmer and drier conditions; tourism promotion and development policies may need to take such risks into account so as not to aggravate existing vulnerabilities.

3.8. Some donor-supported projects implicitly include, or are synergistic with, adaptation

While donor country assistance strategies do not generally pay explicit attention to climate change, some development co-operation projects are synergistic with, or implicitly include, adaptation to climate change. These projects extend beyond social development that might reduce baseline vulnerability through improvements in living standards; rather, they are targeted towards reducing net exposure to specific risks that climate change could aggravate.

In the case study countries, the development co-operation project that is perhaps most directly related to climate change is the Tsho Rolpa risk reduction project in *Nepal*. The Tsho Rolpa is a high-altitude lake formed by the melting and retreat of a glacier as a result of rising temperatures. By 1998, it was the largest and most dangerous glacial lake in Nepal, containing 90-100 million cubic metres of water behind a natural moraine dam. The risk of catastrophic breach of the dam had considerably increased, entailing the possibility of serious damage to settlements and development infrastructure (including a hydropower facility) for 100 km or more downstream. Funded by the Netherlands Development Agency and the Government of Nepal, with additional support from the UK Department for International Development, the project aimed to reduce this risk through a combination of measures, including drainage of the lake and establishment of early warning systems downstream. Substantial drainage of the lake has been accomplished. The phrase "climate change" does not occur in any project documents. Yet, the project clearly accomplished anticipatory adaptation to climate change by considerably reducing the risk of catastrophic flooding, even if continued rising temperatures may cause further lake expansion.

Examples of development co-operation projects aimed at anticipatory adaptation are rare. Somewhat more common are instances where projects that address current climate impacts or related societal vulnerabilities may have synergies with adaptation to climate change.

For example, *Bangladesh* is already vulnerable to coastal flooding, and its vulnerability is projected to increase under climate change, through a combination of factors.[2] Improvements to coastal embankments and decongestion of drainage systems are good adaptations to current vulnerability. Development co-operation activities involving such responses include the ADB-funded Khulna-Jessore Drainage Rehabilitation Project and the Coastal Embankment Rehabilitation Project funded by the World Bank, the European Commission and the Government of Bangladesh. Such projects are also synergistic with adaptation to higher flooding risks under climate change.

In *Egypt*, the productivity and sustainability of water use are already important concerns because of growing demand and potentially decreasing supply as competing demand for the Nile waters grows in upstream countries. Climate change could exacerbate such concerns by increasing evaporation losses and possibly reducing rainfall in the Nile headwater areas. Some development co-operation activities, such as the World Bank Integrated Irrigation Improvement and Management Project, address current water management concerns. Several donors are supporting the intergovernmental Nile Basin Initiative, in which all Nile riparian countries collaborate politically and technically on a range of activities. While climate change is not on its agenda, the initiative could be a useful forum in which to address basin-wide implications of climate change. Another activity is the World Bank/GEF Second Matruh Resource Management Project, which addresses rural poverty. Biophysical risks facing the Matruh region include problems associated with water scarcity, rainfall variability, strong winds and agricultural pests, compounded by low crop diversity. The project recognises the importance of traditional coping mechanisms, such as planting drought-adapted species, tailoring the timing and area of sowing to rainfall, favouring low-input farming and complementing field crops with horticulture, livestock and off-farm economic activities. Some of these, however, such as low-input farming, impede economic progress. Hence, the project builds on and complements existing coping strategies by strengthening local capacity for conservation, rehabilitation and sustainable management of natural resources. It also promotes community development, improved access to services and new income-generating opportunities. All these elements, particularly when tailored to local needs and circumstances, would be synergistic with adaptation to climate change.

4. Concluding remarks

A significant share of core development activities in the case study countries – in both absolute and percentage terms – is in sectors potentially affected by climate change. Moreover, much more official aid goes to activities potentially affected by climate change than to financing of adaptation as part of the climate change regime. Over the long term, the potential for mainstreaming adaptation within core development activities thus appears to be far greater than financing of action on adaptation initiated from within the climate change regime. The exposure of donor portfolios to climate risk is relatively easy to quantify using standardised data on donors, activities and countries. As national development portfolios are also potentially affected by climate change, mainstreaming is needed within national development plans and projects as well.

Assessment of a range of development activities for Bangladesh, Egypt, Fiji, Nepal, Tanzania and Uruguay reveals a fairly nuanced picture in terms of the degree of attention to climate change concerns. On the one hand, considerable progress has been made over the past decade or so with regard to activities specific to climate change, including GHG emission inventories and mitigation measures and climate change impact assessment and adaptation. Institutional mechanisms addressing climate change have been established, and certain countries (such as Tanzania) have developed climate change action plans. Donors have provided financial and technical support for many of these activities. In some cases – as in Bangladesh and Fiji – donors have worked with the national government to better articulate links between responses to climate change and priorities for development. Such initiatives remain largely theoretical at this stage, however, and discernible progress in terms of policy action has been limited.

On the other hand, many core development activities relate to areas that could be affected by climate change. At the level of national governments and donor countries and agencies, such activities include elaboration of high-level strategy documents, establishment of budget priorities, decisions on investments and implementation of infrastructure and capacity-building projects. Most, however, pay little if any explicit attention to climate change, from national development plans and long-term perspectives to country assistance strategies and Poverty Reduction Strategy Papers. At the sectoral

BRIDGE OVER TROUBLED WATERS – ISBN 92-64-01275-3 – © OECD 2005

and project level, the situation is much more nuanced. In isolated instances, ancillary climate benefits are explicitly recognised. An example is the law forming the basis of Uruguay's 1989 Forestry Promotion Policy, which has led to a dramatic increase in carbon sequestration. Anticipatory adaptation to climate change impacts is an implicit focus of the risk reduction project at the Tsho Rolpa glacial lake in Nepal. Beyond such examples directly linked to climate change, several development activities and projects dealing with current climate risks might be synergistic with adaptation to climate change impacts. More generally, a range of social development activities implicitly reduce vulnerability to climate risks.

Notes

1. The CRS database lists almost 5 000 development co-operation activities for the six countries, representing more than USD 15 billion over 1998-2000. Even to examine all project documents would have been beyond the scope of this analysis.
2. These include sea level rise, potentially greater cyclone intensity, higher river flows and increased sediment loads from glacier melt in the Himalayas, and potential intensification of the summer monsoon.

ANNEX 3.A1

DAC Purpose Codes in the Selection of Climate-Affected Projects

DAC code	General sector name	Purpose codes that are included in the selection
110	Education	Not included
120	Health	12250 (infectious disease control)
130	Population	Not included
140	Water supply and sanitation	14000 (general)
		14010 (water resources policy/administration management)
		14015 (water resources protection)
		14020 (water supply and sanitation – large systems)
		14030 (water supply and sanitation – small systems)
		14040 (river development)
		14050 (waste management/disposal)
		14081 (education/training: water supply and sanitation)
150	Government and civil society	15010 (economic and development policy/planning)
160	Other social infrastructure and services	16330 (settlement) and
		16340 (reconstruction relief)
210[1]	Transport and storage	All purpose codes
220	Communications	Not included
230	Energy	23030 (renewable energy)
		23065 (hydro-electric power plants)
		[23067 (solar energy)]
		23068 (wind power)
		23069 (ocean power)
240	Banking and financial services	Not included
250	Business and other services	Not included
310	Agriculture, forestry, fishing	All purpose codes
320	Industry, mining, construction	Not included
330	Trade and tourism	33200 (tourism, general)
		33210 (tourism policy and admin. management)

DAC code	General sector name	Purpose codes that are included in the selection
410	General environment protection	41000 (general environmental protection)
		41010 (environmental policy and management)
		41020 (biosphere protection)
		41030 (biodiversity)
		41040 (site preservation)
		41050 (flood prevention/control)
		41081 (environmental education/training)
		41082 (environmental research)
420	Women in development	Not included
430	Other multi-sector	43030 (urban development)
		43040 (rural development)
510	Structural adjustment	Not included
520[1]	Food aid excluding relief aid	52000 (dev. food aid/food security assist.)
		52010 (food security programmes/food aid)
530	Other general programme and commodity assistance	Not included
600	Action relating to debt	Not included
700[1]	Emergency relief	70000 (emergency assistance, general)
710[1]	Relief food aid	71000 (emergency food aid, general)
		71010 (emergency food aid)
720[1]	Non-food emergency and distress relief	72000 (other emergency and distress relief)
		72010 (emergency/distress relief)
910	Administrative costs of donors	Not included
920	Support to NGOs	Not included
930	Unallocated/unspecified	Not included

1. Sector codes that are excluded in the narrower selection.

ISBN 92-64-01275-3
Bridge Over Troubled Waters
Linking Climate Change and Development

Chapter 4

Climate Change and Natural Resource Management: Key Themes from Case Studies

by
Shardul Agrawala, Simone Gigli, Vivian Raksakulthai, Andreas Hemp, Annett Moehner, Declan Conway, Mohamed El Raey, Ahsan Uddin Ahmed, James Risbey, Walter Baethgen and Daniel Martino

Natural resources and their management form a critical interface between climate change and development. The impacts of climate change can affect the quality and reliability of many of the services natural resources provide. On the other hand, natural resources play an important role in greenhouse gas mitigation and also serve as a first line of defence against climate change. This chapter examines opportunities and trade-offs in integrating climate change considerations in natural resource management in six key systems: the Nepal Himalayas, Mount Kilimanjaro in Tanzania, the Nile in Egypt, the Bangladesh Sundarbans, coastal mangroves in Fiji and the agriculture and forestry sectors in Uruguay. Climate change impacts and adaptation are the primary focus in five cases, while that of Uruguay examines links with mitigation.

1. Introduction

Natural resources and their management form a critical interface between climate change and development. Beyond providing many goods and services that sustain rural livelihoods, promote environmental quality and advance economic development, natural resources serve as a first line of defence against climate change. The impacts of climate change can affect the quality, quantity and reliability of many of the services natural resources provide. This influence, in turn, can affect a range of development objectives, from poverty alleviation to sustained economic growth. Moreover, natural resource management policies play a critical role in climate change mitigation in several ways, from promoting opportunities to sequester carbon through afforestation, to helping reduce GHG emissions through enhanced use of biomass and hydropower.

This chapter examines opportunities and trade-offs in integrating climate change considerations in natural resource management and development objectives in six key systems – the Nepal Himalayas, Mount Kilimanjaro in Tanzania, the Nile in Egypt, the Bangladesh Sundarbans, coastal mangroves in Fiji and the agriculture and forestry sectors in Uruguay. This selection includes several conservation icons – Mount Kilimanjaro and the Sundarbans are UNESCO World Heritage Sites, while the Nile system and the Himalayas are home to several such sites. While the limited number of cases discussed imposes some obvious limits to generalisation, the six case studies do nevertheless encompass a broad range of natural and managed systems, including mountains, water resources, forests, coastal zones and agriculture. Climate change impacts and adaptation are the primary focus in five of the six cases, while that of Uruguay examines links between GHG mitigation and policies for agriculture and forestry. The following sections describe each case in some detail and discuss synergies and trade-offs between climate considerations, on the one hand, and, on the other, priorities related to conservation and economic development. The concluding section highlights cross-cutting issues that emerge from the case studies.

2. Glacier retreat and glacial lake outburst flooding in the Nepal Himalayas

Nepal is a small, landlocked Himalayan kingdom located between the Asian giants China and India. It contains eight of the ten highest mountain peaks in the world, including Mount Everest at 8 848 metres above sea level. It

BRIDGE OVER TROUBLED WATERS – ISBN 92-64-01275-3 – © OECD 2005

also contains regions that are only about 80 metres above sea level. This variation in altitude, from north to south, occurs within only about 200 km. Consequently, Nepal's climate shows extreme spatial variation, from arctic to tropical within a distance comparable to the size of an average grid box of a climate model.

Despite its natural beauty and enormous potential for hydropower and tourism, Nepal is among the poorest countries in the world, with 82.5% of the population living below the international poverty line of USD 2 per day. In fact, about 38% of its population survives on less than USD 1 per day (World Bank, 2002). The population is overwhelmingly rural, and agriculture is the source of livelihood for nearly 81% of the population, contributing 40% of GDP. Most of the agriculture relies on water from the many Himalayan rivers that are fed by the monsoon during the wet season and snow melt during the dry season. Nepal's economy also depends heavily on foreign exchange earned from tourism related to its mountains and other natural resources. In addition, its electricity infrastructure is heavily reliant on hydropower, which provides 91% of the electricity generated.

2.1. *Implications of climate change*

Glaciers and water resources are the principal arteries that feed almost all aspects of Nepal's development. Rivers fed by snow melt and monsoons supply water for the agriculture that supports most of the population, and for hydropower. Nepal's glaciated peaks are a primary catalyst for foreign tourism. Recent climate trends and projected climate change pose significant implications for these critical natural resources. Significant warming has been documented in Nepal since the late 1970s. An important feature of this trend is that higher elevations in the Himalayas have experienced significantly greater warming (Shrestha *et al.*, 1999). Measurements from the other side of the Himalayas, in Tibet, confirm the sensitivity of the highest elevations to warming temperatures (Liu and Chen, 2000).

Greater warming at higher elevations is of particular concern for Nepal, as it exacerbates glacier retreat, and ice melt more generally. While trends in rainfall and river flow are much less robust than the trends observed in temperature and glacier retreat, nevertheless increases in intense rainfall events and the number of consecutive flood days have been documented (Shakya, 2003).

Climate change scenarios for Nepal from multiple climate models, moreover, show consistent patterns of increased warming, with average temperature increases of 1.2 °C projected by 2050 and 3 °C by 2100 (OECD, 2003a). There is also indication that the summer monsoon might intensify, which would be consistent with observed increases in intense summer rainfall.

2.1.1. *Impacts on glacier retreat and river flow*

The observed changes in climate have already had significant impacts in the Nepal Himalayas. Glaciers in Nepal are in a general state of retreat, with measured retreat rates in many cases ranging from 30 to 60 metres per decade (Shrestha and Shrestha, 2004). This retreat is expected to continue, or even accelerate, under projected climate change. Continued glacier retreat and the reduction in winter precipitation also projected under climate change will reduce dry season river flows. This point is critical because the dry season can last as long as nine months and any reduction in flows will have major consequences for agriculture and hydropower generation. On the other hand, any intensification of the summer monsoon – as has already been observed and is projected under climate change – will increase the vulnerability of people and infrastructure to floods and landslides. Landslides not only destroy fields, settlements, forests and roads, but also increase sediment loading in rivers. This in turn can shorten the operating life of water supply and hydropower infrastructure. The combined effect of reduced dry season flows and greater river discharge during the monsoon will be increased variability and reliability of river flows (Shakya, 2003).

2.1.2. *Glacial lake outburst floods*

Glacier retreat and ice melt more generally are also adding considerably to the size and volume of several of the more than 2 000 glacial lakes in Nepal. Glacial lakes form behind unstable ice or moraine dams, as well as beneath or on top of glaciers. Many glacial lakes drain periodically through natural outlets when water reaches a certain level. However, if such drainage is not possible or sufficient, the lakes can breach the dams, producing the catastrophic phenomena known as glacial lake outburst floods or GLOFs.

GLOFs were first observed in Iceland and called *jokulhlaup* (literally "glacier-leap"). They are catastrophic discharges of large volumes of water following the breach of the natural dams that contain glacial lakes. The water rushes downstream in flood waves, with devastating consequences for settlements and infrastructure. GLOFs can be triggered by earthquakes, spontaneous breakage of the moraine dams, avalanches, the collapse of "hanging glaciers" into a lake and other phenomena (Ives, 1986). Climate change and higher temperatures are contributing to a substantial increase in the area and volume of many glacial lakes in Nepal, significantly raising the probability of GLOFs occurring. Table 4.1 presents a list of major GLOFs recorded in Nepal. It shows that many originated across the border in Tibet, underscoring the fact that such events can have transboundary impacts.

The most significant GLOF in Nepal in terms of recorded damage occurred in 1985. A surge of water and debris up to 15 metres high flooded down the Bhote Koshi and Dudh Koshi rivers for 90 km. At its peak, the discharge was

Table 4.1. **GLOFs recorded in Nepal**

	River basin	Name of lake
450 years ago	Seti Khola	Machhapuchhare, Nepal
1935	Sun Koshi	Taraco, Tibet
1964	Arun	Gelaipco, Tibet
1964	Sun Koshi	Zhangzangbo, Tibet
1964	Trishuli	Longda, Tibet
1968	Arun	Ayaco, Tibet
1969	Arun	Ayaco, Tibet
1970	Arun	Ayaco, Tibet
1977	Dudh Koshi	Nare, Tibet
1980	Tamur	Nagmapokhari, Nepal
1981	Sun Koshi	Zhangzangbo, Tibet
1982	Arun	Jinco, Tibet
1985	Dudh Koshi	Dig Tsho, Nepal
1991	Tama Koshi	Chubung, Nepal
1998	Dudh Koshi	Sabai Tsho, Nepal

Source: Shrestha and Shrestha (2004).

2 000 m^3/sec, two to four times greater than the maximum monsoon flood level. The GLOF destroyed the almost-completed Namche Small Hydro Project, which had cost over USD 1 million. Severe erosion destroyed the weir and head race canal where water would flow into the plant (Raksakulthai, 2003). The damage extended 90 km downstream. Fourteen bridges, including new suspension bridges, were destroyed. Erosion, undercutting and destabilisation occurred on long stretches of the main trail to the Mount Everest base camp. When the trail reopened, prices of staple supplies were 50% higher on average (Ives, 1986). While the loss of human life was limited, it could have been high had the GLOF occurred during peak trekking season. The loss of livelihoods was considerable: vast tracts of arable land were rendered unusable along with critical lifelines such as bridges and roads.

An inventory in 2001 found over 3 252 glaciers, 2 323 glacial lakes and 20 potential GLOF sites in Nepal (Mool *et al.*, 2002). In addition to this picture of static risk, site-based monitoring of specific glacial lakes has shown evidence of lake volumes increasing over time. The trend in volume increase correlates well with the observed trends in temperature increase at high altitudes in the Himalayas, discussed earlier. Taken together, the evidence points to a serious hazard closely tied to temperatures rising because of climate change.

2.2. *Adaptation options for GLOF risk and stream flow variability*

Several adaptation strategies are available to cope with GLOF risk and stream flow variability. Some of these responses are already at varying stages of implementation within development projects. The emphasis so far has been much more on engineering solutions than on social measures to reduce vulnerability.

2.2.1. *Siting of hydropower facilities in low-risk locations*

One way to protect hydropower infrastructure from GLOFs is to install such facilities in locations where the threat of GLOFs is low. Documents from the Namche project that was destroyed in 1985 give no indication that any "special attention was paid to the possible occurrence of catastrophic geomorphic events, despite the fact that the project was being sited in one of the highest and most precipitous mountain regions in the world" (Ives, 1986). After the Namche disaster, however, the plant was rebuilt in another location. It has been under continuous operation ever since, and the estimated risk of GLOF at the new location is low.

In general, however, relocating existing facilities is probably not feasible. Siting in low-risk locations may be possible only for new facilities. Even then, questions can arise as to whether generating capacity is lower or transmission costs higher at the alternate site. In fact, a trade-off between risk and profitability may exist: sites that might offer greater hydropower potential and lower generating costs may also be the ones at higher risk. Given the general uncertainty about GLOF risk, investors and energy planners may be reluctant to choose alternate locations when such risk is only one of many factors in choosing a site. Another key barrier to effective incorporation of GLOF risk in project siting is the lack of reliable spatial mapping of lakes that are at risk of breaching. Developing risk maps, however, is not straightforward since GLOFs can travel as far as 200 km downstream. Catchment-wide analyses would be needed to determine vulnerability downstream from hazardous glacial lakes. Furthermore, run-off from GLOFs can pose just as great a risk to hydropower plants by forming large reservoirs that could burst themselves. Indeed, the risk may be even greater, since such reservoirs would be much closer to the plants. An integrated risk management approach is needed, therefore, to supplement satellite-based risk mapping of the lakes themselves.

2.2.2. *Early warning systems*

Given the catastrophic nature of GLOFs, early warning systems need to be established downstream from hazardous glacial lakes. While many disasters can be forecast with some precision (*e.g.* those related to El Niño), no such capability exists for GLOFs. Nevertheless, once an event is under way, timely warning can save lives downstream. The effect of early warning systems is likely to be significant only far downstream, however, given the pace of GLOFs. Moreover, early warning cannot save infrastructure critical to local livelihoods, such as hydropower facilities, bridges and roads. A key additional constraint is the high set-up costs of automated GLOF early warning systems, estimated at USD 1 million per river basin. Adequate warning systems would represent a major investment for a poor country like Nepal, since several of its river basins are at risk, and the risk will increase with rising temperatures.

2.2.3. *Micro- versus storage hydro*

Climate change entails two sets of impacts on hydropower generation that may require conflicting adaptation responses. On the one hand, increased GLOF risk would argue for increasing the number of dispersed micro-hydro facilities in small river basins. This option would diversify the risk and limit the overall loss of infrastructure and generation capacity if a facility were to be affected by a GLOF. Micro-hydropower has the potential to meet much of the rural demand for energy in Nepal. Installation of micro- and small hydro facilities is consistent with Nepal's development priorities and is being encouraged by both the government and donors. Climate change thus might be an additional reason to promote a strategy that is already being implemented for reasons of economic development.

On the other hand, as has been noted, climate change is likely to contribute to increased seasonal variability of river flows and reduced dry season flows. Reliability of river flows and baseline minimum flows during the dry season are critical for hydropower generation. Since smaller rivers are expected to show greater variability, adaptation to these impacts would require building hydropower facilities in larger river basins. Reduction in dry season flows, moreover, might militate for storage hydro rather than the currently prevalent "run of river" projects. In terms of GLOF risk, however, storing large volumes of water might be a maladaptation as it could cause a "double whammy" effect if the reservoir were breached following a GLOF. This conundrum highlights an important point: adaptation responses may not always be internally consistent, or, for that matter, coherent with other social or development priorities. Conflicts may arise, requiring careful consideration of the trade-offs involved.

2.2.4. *Incorporation of GLOF and stream flow variability considerations in project design*

While damage to downstream hydropower infrastructure in the event of a GLOF cannot be avoided, it can be limited through design measures. For example, the powerhouse can be placed underground, the tailrace can be protected and the reservoir design can take into account the risk of excess sediment deposition (Shrestha and Shrestha, 2004).

Adaptation to less certain impacts can also be incorporated through greater design flexibility. Hydropower generation already involves several mechanisms to cope with stream flow fluctuations resulting from current seasonal and climate variability. For example, a plant may have three intake channels and turbines for the peak run-off (monsoon) season, one or more of which can be shut during the dry season. This allows the plant to generate electricity more efficiently and without incurring losses from excess capacity. Consideration of this option should include analysis of the economic benefit of designing hydropower plants to accommodate the possibility of lower capacity in the future (for example, in 25 years). The current design standard for small hydropower assumes an average lifespan of 50 years, and most investors expect a return on their investment within seven years. This situation highlights another challenge in the medium to long term, given the considerable uncertainty regarding the magnitude and timing of many climate change impacts.

2.2.5. *Direct reduction of GLOF risk*

Another type of adaptation response to GLOF hazards involves physically reducing flood risks from glacial lakes. Methods include siphoning or pumping water out of dangerous lakes, cutting a drainage channel, and taking flood control measures downstream (Rana *et al.*, 2000). Some of these measures may have ancillary benefits. Siphoned water could, for example, be used to supplement dry season flows, maintain adequate water levels in downstream ecosystems to protect valuable fish stocks, help meet local demand for water supply and even provide recreational facilities.

Such measures, however, have disadvantages. Pumping, for example, is expensive because the remote, high-altitude locations mean heavy equipment must be flown in by helicopter. Building a siphon could breach the moraine dam and risk triggering a GLOF. Disadvantages notwithstanding, in one instance in Nepal integrated direct risk reduction has been undertaken (Box 4.1). The measures in the Tsho Rolpa Risk Reduction Project may need to be repeated as lake levels rise again with further ice melt, however. Furthermore, perhaps a dozen or more other lakes in Nepal alone pose considerable risk, but to replicate such measures is beyond the capacity of the government and donors, given the cost of lowering the level of just one lake. Nevertheless, the potential cost of damage from a GLOF could be much greater, in terms of lost lives and communities, development setbacks and forgone energy generation.

2.3. *Implications for donors and transboundary co-ordination*

This analysis reveals that the need to integrate considerations of climate change in development activity is acute in Nepal. Glacier retreat and GLOF are both closely tied to rising temperatures and are already affecting people and development infrastructure. Adaptation to these impacts and other projected changes, such as reduced dry season rainfall and stream flow, is therefore critical in the case of Nepal. A discussion of initiatives already under way reveals a rather complex mosaic of choices that frequently have high upfront

 BRIDGE OVER TROUBLED WATERS – ISBN 92-64-01275-3 – © OECD 2005

Box 4.1. **Tsho Rolpa Risk Reduction Project**

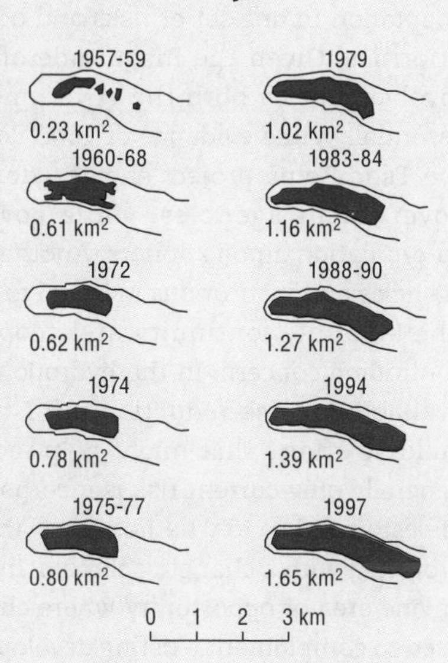

One of the most dangerous glacial lakes in Nepal is Tsho Rolpa, at an altitude of about 5 000 metres. Its size increased from 0.23 km^2 in 1957/58 to 1.65 km^2 in 1997. The Tsho Rolpa project is an important example of collaborative anticipatory planning by government, donors and experts in GLOF mitigation.

Tsho Rolpa contained about 90-100 million m^3 of water in 1997, a hazard that called for urgent attention. A moraine dam 150 metres tall held the lake. If it were breached, one-third or more of the water could flood downstream. Among other considerations, this posed a major risk to the 60 MW Khimti hydropower plant, which was under construction downstream. These concerns spurred the Government of Nepal to initiate a project in 1998, with the support of the Netherlands Development Agency, to lower the level of the lake through drainage. An expert group recommended that, to reduce the risk of a GLOF, the lake should be lowered three metres by cutting a channel in the moraine. A gate was constructed to allow for controlled release of water. Meanwhile, with a World Bank loan, an early warning system was established in 19 villages downstream on the Bhote/Tama Koshi River in case a Tsho Rolpa GLOF should occur despite these efforts. Local villagers were actively involved in the design of the system, and drills are carried out periodically. Construction on the four-year Tsho Rolpa project was completed in December 2002 at a cost of USD 3.21 million: USD 2.98 million from the Netherlands and USD 0.23 million from Nepal. By June 2002, the goal of lowering the level by three metres had been achieved, reducing the risk of a GLOF by 20%. Complete prevention of a GLOF would require further drainage to lower the lake level by perhaps as much as 17 metres. Expert groups are undertaking further studies. Clearly, mitigating GLOF risk involves substantial costs and is time consuming.

Source: OECD (2003a); figure from Department of Hydrology and Meteorology (2005).

costs, involve making decisions under significant uncertainty and often entail complex trade-offs, such as between risk reduction and profitability or between adaptation to one set of risks and other adaptation measures or development priorities. Given the magnitude of the costs of many of these measures, involvement by both the government and international donors has been essential. While evidence of good donor-government collaboration exists – as the Tsho Rolpa project shows – donors cite a lack of co-ordination among government agencies, while government agencies point to a lack of co-ordination among donors. Another constraint is the limited capacity of host agencies and institutions in Nepal to field multiple and diverse donor requests. The amount, continuity and scope of project funding also constitute a continuing concern. In the hydropower sector, funding has been more readily available for risk reduction infrastructure than for training and capacity-building efforts that might help reduce societal vulnerability. Furthermore, generally only current risk is incorporated in project planning. The evidence is at best mixed as to whether plans and projects incorporate the increase in risk projected with a changing climate. Incorporation of such long-term risks might be one area of opportunity where climate change funds and projects could be used to complement existing development funding by focusing on training and capacity building, as well as longer-term risk and vulnerability reduction.

Finally, there is an important transboundary or regional dimension to climate change impacts and responses. Many catastrophic GLOF events in Nepal originate in Tibet. Decisions about water resource management or hydropower generation in Nepal affect neighbouring countries such as Bangladesh and India. In addition to national discourses on links between climate change and development, discussions may be needed at a regional level to formulate co-ordinated strategies.

3. Ice cap melt and forest fire risk on Mount Kilimanjaro

Mount Kilimanjaro is a huge stratovolcano located in East Africa, in northern Tanzania on the border with Kenya. It is composed of three peaks, Kibo, Mawenzi and Shira, with Kibo being the highest at 5 895 metres. Kilimanjaro is the highest mountain in Africa. Its name derives from the Swahili *Kilima Njaro*, "shining mountain", a reference to its legendary ice cap. Today permanent ice exists only on Kibo, covering an area of 2.6 km^2. Yet, the distribution of moraines reaching down to an altitude of 3 000 metres indicates that the ice formerly covered a much greater area. The retreat and eventual loss of the ice cap has made the mountain a prominent symbol of the impacts of global climate change. In addition, Mount Kilimanjaro is a biodiversity hot spot, with nearly 3 000 plant species. It provides a range of critical ecosystem services to the more than 1 million local people who depend on it for their livelihoods, as well as to a broader region that depends

BRIDGE OVER TROUBLED WATERS – ISBN 92-64-01275-3 – © OECD 2005

on water resources originating on the mountain. Many of these ecosystem services have also been critically affected by long-term trends in climate, along with human pressures.

3.1. *Current climate and climatic changes*

Mount Kilimanjaro has a typical equatorial daytime climate. The driest period is from July to October, while April and May are the wettest months. Rainfall and temperature vary with altitude and exposure to the dominant winds blowing from the Indian Ocean. Annual rainfall reaches a maximum of around 3 000 mm at an altitude of 2 100 metres on the central southern slope and declines with increasing altitude. The northern slopes, on the leeward side of the mountain, receive much less rain. The mean annual temperature is 23.4 °C at the base of the mountain and falls about 0.6 °C per hundred metres of increase in altitude, reaching 5 °C at 4 000 metres and –7.1 °C on top of Kibo.

Although it is not possible to infer temperature trends at different altitudes from the available data, a distinct overall warming trend has been observed for most of the period from 1950 to the present. Average temperatures in the Kilimanjaro region rose in 1951-60, were stable in 1960-81 and increased again in 1981-95 (Hay *et al.*, 2002). The record of temperatures from 1976 to 2000 in the neighbouring Amboseli region reveals that mean daily maximum temperature rose at a rate of over 2 °C per decade (Altmann *et al.*, 2002), significantly higher than globally averaged warming, with the increase being greatest during the hottest months, February and March. Rainfall data from Mount Kilimanjaro indicate a steady decline in precipitation over the past century (Hemp, 2005). In addition, the number of months with less than 30 mm of rainfall appears to have increased; the number of wet months is unchanged. While natural climate variability and land use changes have likely contributed to these recent developments, the warming trend of the past few decades is consistent with human-induced climate change.

Either declining precipitation or increasing temperature, or both, could contribute to increases in glacier melting and fire risk. These increases have been empirically observed in recent decades. Their implications are discussed below.

3.2. *Potential impacts of climatic changes*

3.2.1. *Glacier retreat*

The ice cap on Mount Kilimanjaro has been retreating since the end of the Little Ice Age around 1850. This retreat has been driven by natural climatic shifts, particularly a decline in regional precipitation of the order of 150 mm, along with changes in cloudiness during the last quarter of the 19th century. The retreat in the ice cap appears to have accelerated because of warming observed in the second half of the 20th century. Between 1962 and 2000, Kilimanjaro lost about 55% of its glaciers. There is general consensus that by 2020 the ice cap will disappear for the first time in over 11 000 years. The symbolism of this loss notwithstanding, it is important to note that its impact on natural and human systems will be very limited. The glaciers of Kibo cover an area equivalent to 0.2% of the forest belt area on Kilimanjaro. Only two rivers are directly linked to the glaciers, by very small streams, and 90% of the precipitation is trapped by the forest belt. Even after the glaciers have melted, precipitation on Kibo will feed springs and rivers, although not so continuously and to a much lesser degree.

Therefore, it is very unlikely that the loss of the glaciers will have a major impact on the hydrology of the mountain. Observations of dry river beds are not necessarily an indicator of long-term climatic changes or the impact of shrinking glaciers. Dried-out rivers in some areas are much more likely to be the result of forest destruction or increasing water demand from a rapidly growing population. Water diversion has quadrupled in certain areas during the past 40 years (Sarmett and Faraji, 1991). Nor is it certain that the loss of the ice cap will affect tourism revenue. Certainly, Kilimanjaro will lose part of its beauty along with its glaciers. It will remain the highest mountain in Africa, however, and this will probably prove to be attraction enough. A significant decline in tourist numbers, then, is also unlikely.

3.2.2. *Increased fire risk*

A less publicised yet potentially far more significant impact of climate change on Mount Kilimanjaro is the intensification of fire risk and attendant consequences for biodiversity and ecosystem services. The mountain has a series of rich and diverse vegetation zones, stratified by altitude. Coffee and banana plantations at the base give way to montane forests at around 1 700 metres, with a transition to a subalpine belt at around 2 800 metres. In the subalpine area, a cloud forest zone with *Erica excelsa* gives way to scrubland, then to bush and cushion vegetation, with increasing altitude. Human action is the primary cause of forest fires in the submontane and montane forest belt. In the upper reaches where human interference is quite limited, however, climatic conditions play an increasingly important role in determining fire risk.

In theory, rising temperatures should result in upward migration of vegetation zones. This effect, however, has been offset by the intensification of fire risk resulting from the warmer and drier climatic conditions. As Figure 4.1 shows, recurrent fires result in progressive degeneration of the vegetation in the subalpine area where shrub replaces forest. Continuously high frequency of fires further harms this bush vegetation, the ultimate result being cushion vegetation.

BRIDGE OVER TROUBLED WATERS – ISBN 92-64-01275-3 – © OECD 2005

Figure 4.1. **Forest succession after continued fires**

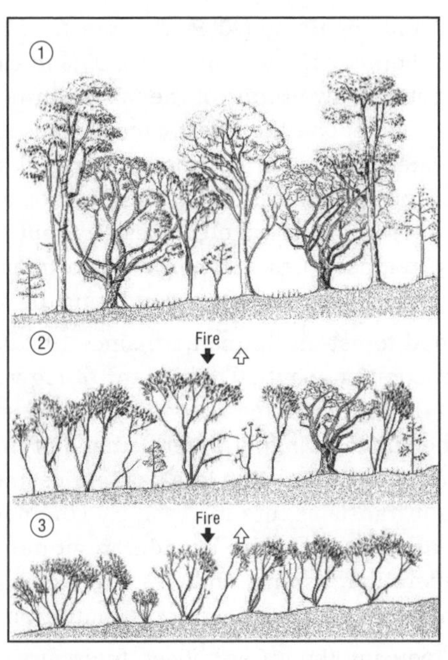

Source: Hemp (2003).

Consequently, climatic changes have actually pushed the upper forest line *downward* on Mount Kilimanjaro from the subalpine to lower and moister regions. A comparison of satellite images from 1976 and 2000 (see Plates 10 and 11, page 129) reveals enormous changes in the upper vegetation zones. In 1976, *Erica* forest covered 187 km^2 – nearly six times the current area of 32 km^2 – and extended in many places up to 3 800 metres. The loss of 155 km^2 represents a 15% reduction in forest cover due to fire since 1976.

3.3. Socio-economic impact of increasing fire intensity

The loss to fire since 1976 of over 150 km^2 of forest, mostly *Erica* cloud forest, in the upper reaches of Kilimanjaro has seriously disturbed the water balance of the entire mountain. In cloud forests, a dense epiphytic layer absorbs about one-third of the rainfall (Pócs, 1976). Destruction of cloud forest reduces the function of the forest belt as a water filter and reservoir. Instead of remaining in the thick epiphytic biomass, humus and upper soil of the forest and percolating slowly into aquifers, rainwater flows off quickly on the surface to rivers, eroding the soil and increasing the danger of flooding on the foothills. In addition to filtering and storing water, cloud forests tend to collect water through fog interception – water droplets are blown by the wind against the vegetation, where they coalesce to form large drops that run off and fall to the ground.

According to a study by Hemp (2005), the forests of Mount Kilimanjaro receive nearly 1.6 billion m^3 of water annually, 95% in the form of rainfall and 5% by fog interception. Two-thirds of the rainwater returns to the air by evapo-transpiration, whereupon the forest canopy traps it as humidity. The forest thus stores water not only in its biomass and soil but even in the air around it. This mechanism enhances the forest's function as a water reservoir regulating the outflow patterns of watercourses. Without a permanent cloud cover over the forest, temperatures, and hence evapo-transpiration, would be much higher and no rain showers would occur during the dry season.

The loss of cloud forest on Kilimanjaro since 1976 has resulted in an average annual reduction of about 20 million m^3 of fog water. This number represents over 25% of the annual fog water input of the entire forest belt, or the equivalent of the annual drinking water demand of the 1 million people living on Kilimanjaro (Hemp, 2003). The loss has serious implications for local livelihoods because it translates into increasingly common water shortages during the dry season. It also has broader regional implications for hydroelectricity generation, fishing, and rice and sugarcane cultivation in areas that depend at least partially on water originating on Kilimanjaro.

In addition to reducing the water budget, forest fires destroy trees that are a source of large amounts of precious wood. They also destroy flower trees for bees, and plants that serve medicinal purposes or are used as forage by members of the Chagga indigenous group. Repeated burning also modifies the soil nutrient balance.

3.4. Climate risk in perspective: shrinking glaciers versus increased fire risk

The 2.6 km^2 of glaciers remaining on Mount Kilimanjaro constitute water volume of about 72 million m^3. Most of this water is not available to the lowlands because the glacier ablation occurs by sublimation, and the melt water that remains evaporates immediately (Kaser *et al.*, 2004). If one-fourth (18 million m^3) of the glacier water were to reach rivers, the output would average about 0.9 million m^3 a year up to 2020, when, if the forecasts are borne out, the glaciers will be gone.

Meanwhile, Kilimanjaro is receiving 20 million m^3 less water each year because of forest depletion and vegetation changes due to forest fires since 1976. If current trends in climatic changes, fire frequency and destructive human influence continue, most of the remaining subalpine *Erica* forest could disappear within the next few years. That would mean the loss of Kilimanjaro's most effective water catchment area, as fog interception is of highest importance in the

 BRIDGE OVER TROUBLED WATERS – ISBN 92-64-01275-3 – © OECD 2005

Erica forest. The result, the loss of about 35 million $m^3$of fog water each year, is an impact of far greater magnitude than the loss of the ice cap. Still, the melting glaciers are certainly an alarming indicator of severe environmental changes on Mount Kilimanjaro.

3.5. Policy responses for Kilimanjaro

As in Nepal, some impacts experienced on Mount Kilimanjaro may have been inevitable because of long-term climatic trends. Little can be done to reverse or even delay the loss of the ice cap. Better management practices, however, can go a long way towards: *i)* reversing the increase in fire risk resulting from the warmer and drier climate; *ii)* fighting forest fires once they start; and *iii)* reversing the loss in forest cover. Climate change only adds to the urgency of fire prevention and control and of forest conservation.

The natural montane forest on Mount Kilimanjaro was given protected status in the early 20th century. Cutting of indigenous trees continued nonetheless to increase until 1984, when the severity of the forest destruction led to a presidential order banning all harvesting from the catchment forests on Kilimanjaro. Even these restrictions did not prevent encroachment from continuing. Yet, general awareness of the need to protect the natural resources of the mountain (especially its forests) is high among local people, government agencies and non-governmental institutions that have run reforestation projects.

Various projects and programmes have aimed at enhancing forest management and increasing conservation and afforestation. A catchment forestry project, for example, focuses on improving management of the forest reserve by involving local communities. Villages adjacent to the catchment forest are now responsible for monitoring encroachment into the forest. Village conservation committees are charged with establishing tree nurseries, organising forest patrols, mobilising firefighters and controlling entry into the forest by issuing permits. Since most areas heavily affected by fires are located inside Kilimanjaro National Park, effective park management is a key to reducing fire risk on the mountain. No large-scale effort has been made, however, and most projects initiated by the forest department to protect indigenous forest have failed, as remote sensing surveys show. A recent decision to incorporate the whole forest reserve within the protection boundary of the national park may prove to be a useful first step towards resolving these problems.

Establishing open strips as fire breaks seems generally unsuitable for the steep, inaccessible slopes of Kilimanjaro. This approach might work, however, on the south-eastern slopes where a plateau with moorland vegetation occurs at the fringe of the forest. In this area, grassland fires affecting the bordering forests are common, and open strips to prevent the fires from spreading into the forest could be effective. At the lower forest boundary, fire lines could be reactivated and cleared before the dry season. Meanwhile, employing well-trained foresters to start reforestation projects and educate local foresters would seem to be a must. The choice of tree species is important: riverine areas, in particular, could be replanted with indigenous trees.

Another need is better forest fire control for Kilimanjaro National Park, including early warning systems. Fire observation points could be established on higher hills or ridges near ranger posts or tourist areas, for instance. One or two small airplanes could significantly improve the fire-fighting potential. Fire control funding, now provided by government and donors, could be increased considerably if part of the revenue of Kilimanjaro National Park (currently several million dollars per year) were earmarked for conservation and fire prevention.

Beyond such piecemeal solutions, a better understanding of the livelihood needs of local people is critical if they are to be engaged more fully in conservation and fire prevention efforts. For example, at one point fires were banned in mountaineering camps, but the measure did not have the desired effect because most fires were being lit not by climbers but by honey collectors and poachers. A more sustainable solution would be to identify viable livelihoods that reduce human pressures on the forest. Creative solutions to boost local incomes, such as incentives for switching to specialty coffee production, could be part of a package of responses to help reduce pressure from activities such as logging and honey collection. Thus a critical need is to develop a comprehensive and holistic development plan, focusing on fire risk and forest destruction while also identifying conservation strategies to assure long-term sustainability of the valuable resources of the Kilimanjaro ecosystem.

4. Climate change and Nile water availability in Egypt

The Nile is the longest river in the world, flowing northward for about 6 700 km from East Africa to the Mediterranean. It is formed by four major tributaries: the White Nile, the Blue Nile, Atbara and Sobat. The White Nile originates in Lake Victoria in Central Africa. The Blue Nile, Atbara and Sobat have their sources in the Ethiopian highlands. These three together account for over two-thirds of the total flow of the Nile, with the Blue Nile alone contributing over half of the total.

Ten countries share the Nile Basin, and the river is the principal source of water for three of them. Cradle of the Egyptian civilization over 5 000 years ago, it has been Egypt's lifeline ever since. Egypt does not contribute significantly to the Nile flow but relies on it for 95% of its freshwater needs, particularly for irrigation. While the large, rectangular country covers about 1 million km^2, most of the population, land use and economic activity are

BRIDGE OVER TROUBLED WATERS – ISBN 92-64-01275-3 – © OECD 2005

concentrated in a narrow strip along the Nile and on the coast around the Nile delta. Plates 12-14 (see page 130) illustrates the critical dependence on this narrow lifeline.

A key implication of Egypt's dependence on the Nile is that any adverse trends with regard to reliability of water supplies from the river have a critical impact not just on particular communities or sectors but on the welfare of the entire country.

4.1. *Historical variability in Nile flows*

The humid highlands of East Africa and Ethiopia where the Nile tributaries rise do not experience high variability in rainfall from year to year. Considerable variations have been observed from one decade to another, however. These, which are due to natural shifts in the climate system, have had significant consequences for Nile flows. Between 1900 and 1997, Blue Nile flows ranged from 20 km^3 per year to four times that much. In the 1960s, the level of Lake Victoria, which feeds the White Nile, rose by over two metres, and it has steadily declined ever since (Conway, 2002). Flows of the main river entering Egypt, which integrated both these effects, have generally declined since 1900 (Figure 4.2).

Rainfall variability within Egypt itself is almost inconsequential in this context: the country receives very little rain, and its agriculture is irrigated, not rain-fed. Egypt has always fostered its development by harnessing the Nile waters, however, and variability in Nile flows has had widespread impacts. The annual Blue Nile flood, in which the flow quadrupled or quintupled between July and October, was virtually eliminated when the Aswan High Dam was completed in 1972. Behind the dam, Lake Nasser has a year's worth of storage capacity to help cope with periodic droughts. Thus, since completion of the dam, Egypt has become reasonably well adapted to current climate variability, though the country remains vulnerable to multi-year droughts.

4.2. *Implications of climate change*

Climate change can affect Nile flows in Egypt via: *i)* changes in rainfall at the headwaters; *ii)* rises in evaporation losses as a result of temperature increase and *iii)* changes in water demand (due to a combination of climatic and non-climatic factors) that in turn affect water availability.

The implications of climate change for rainfall in the Nile headwater areas are highly uncertain: the results of various climate models diverge considerably. Projections of flows in the Blue Nile (which, as has been noted, contributes roughly half the total Nile flow) differ considerably for both summer and winter. In the White Nile system, climate models indicate a small to large increase in rainfall during December-February but are not consistent even in terms of whether rainfall increases or decreases during June-August

Figure 4.2. **Variation in Nile flows and in the level of Lake Victoria**

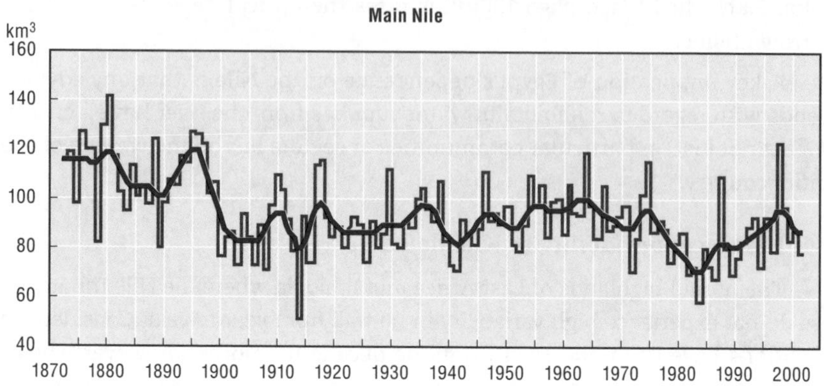

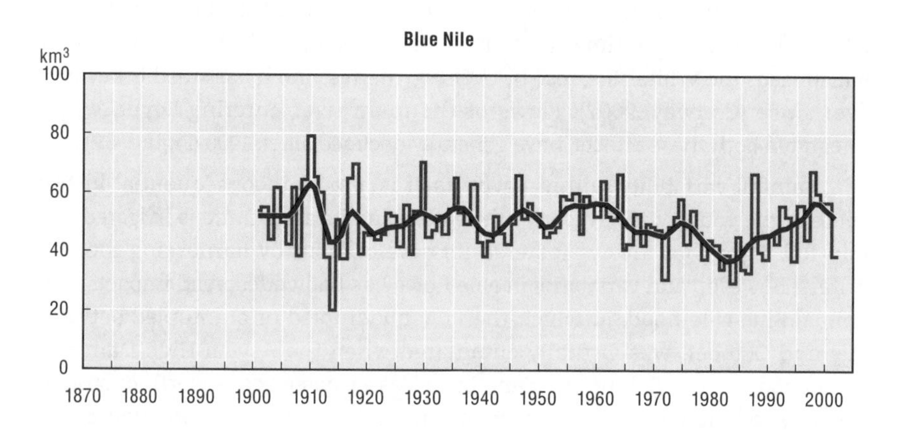

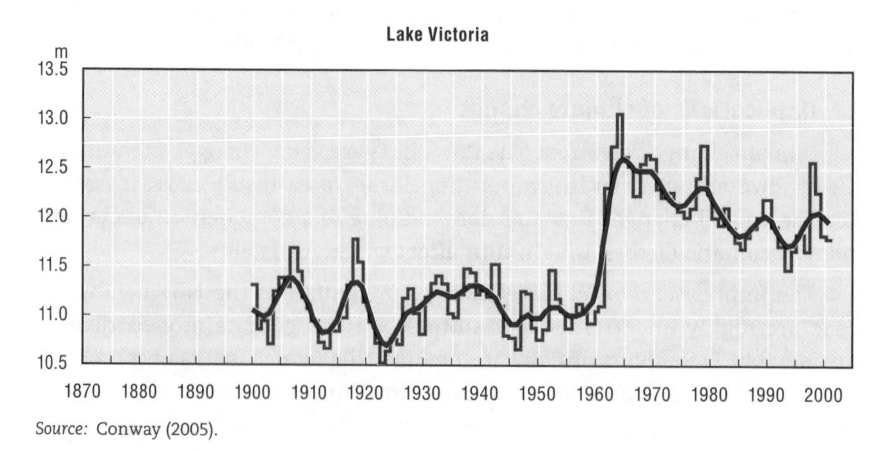

Source: Conway (2005).

(Conway, 2005). The overall result is a wide range of projections concerning the effects of climate change on Nile flows. Some examples suggest a range from a 77% decline to a 30% increase, depending upon the climate models used (Strzepek *et al.*, 1995).

Confidence in the magnitude of observed warming over the past few decades, however, is strong. Climate model projections that this warming will continue and even accelerate also appear robust. A comparison of results from eight recent climate models for Egypt consistently show temperature rising by about 1 °C by 2030, 1.4 °C by 2050 and 2.4 °C by 2100 – greater than the globally averaged projections (OECD, 2004a). Increases of this magnitude throughout the Nile Basin could reduce the levels and outflows of the East African lakes that feed the Nile and exacerbate the already high evaporation losses as the river flows through the semi-arid regions of northern Sudan and Egypt. Evaporation losses already exceed rainfall when the White and Blue Nile enter Sudan. Roughly half the inflows to the Sudd wetland system in southern Sudan are lost to evaporation. Evaporation losses are also critical when the Nile enters and flows through Egypt. At the Aswan High Dam alone, annual evaporation losses total roughly 10 billion m^3.

In addition to affecting evaporation losses from the Nile, higher temperatures are also likely to reduce yields and water use efficiency for key crops such as maize and wheat, and increase water demand for residential and other uses. Population growth and other demographic and development trends will compound any climate-related water stress.

4.3. Nile water allocation and implications for Egypt

The potential vulnerability of the Nile flows in Egypt could be seriously exacerbated should climate impacts be accompanied by a reduction in the allocation of Nile waters to Egypt. The current water allocation stems from the 1929 Nile water agreement between Egypt and Britain, then representing Kenya, Tanzania, Sudan and Uganda. Co-operation to build the Aswan High Dam led to a 1959 follow-up agreement between Egypt and Sudan for "the full utilization of the Nile waters". It allocated 75% (55.5 billion m^3) of the water to Egypt annually and 25% (18.5 billion m^3) to Sudan. The upper riparian countries, however, have disputed provisions of the 1929 and 1959 agreements. Any reallocation resulting from these or other issues could imply a reduction in the Egyptian share of the Nile waters.

Thus, even though Egypt is well adapted to current climate variability, the combination of three factors could increase its future vulnerability: the prospect of a potential reallocation of Nile water; the dependence of Egypt on the Nile combined with growing demand for water and electricity; and potentially negative climate change impacts (notably a reduction in stream flow due to increased evaporation and possibly reduced precipitation). Therefore, measures allowing Egypt to adapt to potential medium- to long-term reductions in water availability need to be formulated, and integrated into the development process.

4.4. *Towards mainstreaming of adaptation responses*

The range of adaptation options that Egypt could use to cope with potential water stress over the medium to long term should theoretically encompass the global, regional and basin-level factors that drive water availability, water use or both. Figure 4.3 shows these factors schematically.

Figure 4.3. **Multi-scale drivers affecting Nile water availability in Egypt**

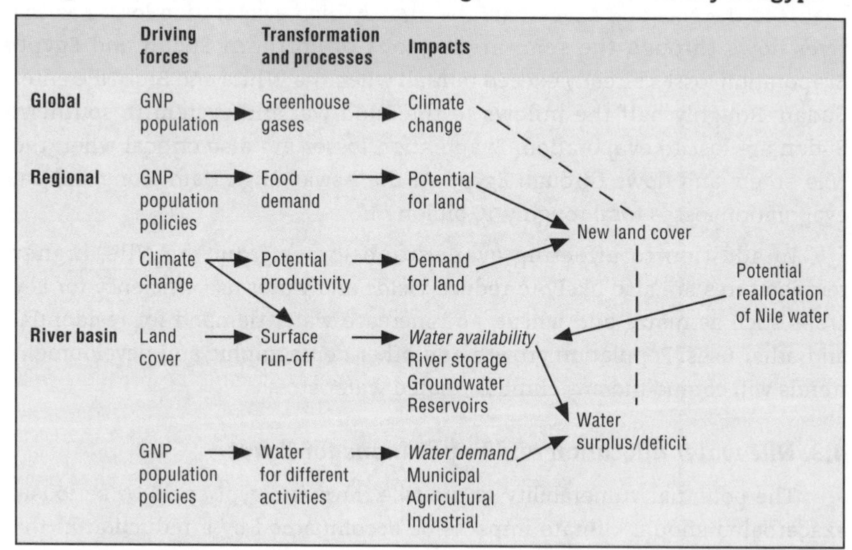

Source: Adapted from Conway *et al.* (1996).

Adaptation in the water resource sector in Egypt is closely intertwined with the development choices and paths of the country and the Nile region. Any changes in water supply due to climate change will occur alongside the certainty of demographic trends and the potential for abstraction by upstream riparian countries. Egypt thus already faces massive water management challenges.

Water resources in Egypt satisfy demand for irrigation, municipal/household water systems, industrial use and electricity generation. Water is also needed to assure barrage safety and navigation efficiency on the Nile and its branches. Adaptation on the demand side requires minimising consumption of water and optimising economic return per unit volume.

Egypt has commissioned a series of strategic water assessments since the 1970s. They contain extensive quantitative measurements and estimates of current and projected water supply and demand in the basin and within Egypt. The studies indicate that Egypt still has some surplus water in the system. Wichelns (2002) quotes estimates of 63.7 km^3 total annual supply and 61.7 km^3 total annual use. Nationally, then, water is not yet a limiting resource. Another study identifies four measures to increase slack within the system: reliance on "virtual water" (*i.e.* importing, rather than producing, water-intensive food products and other goods), economic measures to improve water use efficiency, rainwater harvesting, and drainage and water conservation projects upstream in southern Sudan (Waterbury, 2002). Each of these measures faces several constraints. Reliance on virtual water has economic and political implications, and it is widely held that national food security remains an important political goal. Important socio-economic and political considerations are associated with introducing water use efficiency and other demand-side management measures in Egypt. Many supply-side measures such as rainwater harvesting are limited by the size and accessibility of the resource, and conservation programmes in Sudan (which affect water supply in Egypt) face implementation difficulties. Furthermore, with the population of Egypt growing by a million roughly every nine months, any existing surplus is dwindling fast, and the main opportunities to ease the pressure are not without cost.

The storage capacity provided by the Aswan High Dam insulates Egypt fairly well against the effects of inter-annual variability in Nile flows. The country remains vulnerable, however, to inter-decadal variability. A prolonged dry spell in the 1980s, for example, led to the country facing the prospect of a major water shortage by the summer of 1988. This prospect, in turn, caused the government to initiate policy responses and planning so as to cope with similar situations in the future. Although storage was sufficient to meet established releases, the government made emergency plans in 1988 to counter further worsening of drought conditions. These included improvement in regulations so as to reduce annual releases from Lake Nasser, extension of the winter shut-down of the irrigation system, reduction of the area under rice and improvement of the Nile's navigable channel to keep the water level high enough to allow withdrawal for irrigation. The drought also prompted establishment of a department responsible for monitoring, forecasting and simulating Nile flows (Abu-Zeid and Abdel-Dayem, 1992). These responses are synergistic with adaptation to potential reductions in water supply over the medium to long term. However, the urgency of the situation had declined by the mid-1990s as Nile flows increased and the level of Lake Nasser recovered: in 1997, both reached their highest points since the dam was completed.

Prolonged dry periods require additional response, primarily in terms of contingency planning to limit reservoir releases in a situation of reduced storage and inflows. Measures to deal with prolonged high flows and excess storage must also be considered. Regular reviewing and updating of drought responses, as well as research on improved long-term forecasting, will enhance Egypt's ability to cope with prolonged drought. Climate change may bring more prolonged and even permanent changes in Nile flows, beyond the capacity of current measures to deal with variability, requiring structural changes in management strategies. Timescales are relevant in this respect. Given the magnitude of inter-decadal variability in Egypt, it could be decades before any clear change in flow regime becomes apparent, while it might be beneficial to take certain planning and management steps now.

Finally, dialogue and co-operation among the Nile Basin states are needed to address technical issues (such as data sharing) and more political and sensitive issues (such as water allocation). A co-operative mechanism would be in the interest of all Nile riparian countries, as it would reduce the risk of uncertainty and surprise. It would also promote exploration of adaptation options involving multiple riparian countries in which water consumption is linked to trade in water-intensive commodities such as hydroelectricity and certain food products. While technical co-operation among several basin countries has existed for some time, more comprehensive co-operative agreements were forged only in the 1990s with the active involvement of several donor agencies. This effort eventually led to the establishment in May 1999 of the Nile Basin Initiative (NBI), which now counts all ten basin countries as members. The NBI is structured around two complementary initiatives, one top-down and the other bottom-up: the Shared Vision Program and Subsidiary Action Programs. The objective of the Shared Vision Program is to create an enabling environment for investment and action. Within this basin-wide framework, Subsidiary Action Programs will devise and implement development projects at sub-basin level involving two or more countries, while national and subnational initiatives are left to individual countries. Subsidiary Action Programs encompass a range of linked water resource and economic development issues, including irrigation and drainage, fisheries, hydropower development and pooling, sustainable management of wetlands and watersheds, regional energy networks including grid interconnection, regional transport and communication, trade promotion and disaster forecasting and management. While it is too early to assess the effectiveness of the NBI, the initiative marks an important beginning in terms of providing a co-operative forum in which the water needs and development aspirations of all Nile Basin countries may be reconciled. Climate concerns do not yet figure in NBI activities, but could usefully be incorporated, given their potential significance for all Nile Basin countries.

5. Climate change and coastal mangroves in Bangladesh and Fiji

Mangroves are taxonomically diverse assemblages of plants (including trees, shrubs, ferns and palms) found along sheltered river banks and coastlines in the tropics and subtropics. They frequently occur in conjunction with coral reefs and

BRIDGE OVER TROUBLED WATERS – ISBN 92-64-01275-3 – © OECD 2005

Box 4.2. **Benefits and services of mangrove ecosystems**

Direct use values	Indirect use values	Non-use values
• Forestry products (timber, fuel wood, charcoal, housing material, poles, shingles)	• Flood and erosion control	• Biodiversity conservation
• On-site fishery products (crabs, fish)	• Shoreline stabilisation	• Carbon sequestration
• Support of off-site fisheries (shrimp, fish)	• Storm protection	
• Aquaculture products (shrimp, fish)	• Nutrient sediment trapping	
• Other products, such as food, medicinal plants, honey	• Habitat and nursery provision	
• Recreation and tourism		

seagrass beds as part of a wider coastal ecosystem that has been termed "seascape". Mangroves are among the most productive ecosystems in the world, and provide a wide range of services (Box 4.2). The dense tree foliage allows a high level of primary productivity, which in turn supports a complex food web of terrestrial and marine life. Mangroves provide nutrients and habitat for juvenile fish, crabs and shrimp, and host a wide variety of birds and other wildlife. They also help maintain coastal region integrity by protecting against floods, coastal erosion and storm surges. As such, they are considered a critical "no regrets" adaptation to the impacts of sea level rise and storm surges. At the same time, mangroves are vulnerable to climate change impacts, particularly sea level rise and changes to minimum winter water temperature. They are also threatened already in many parts of the world from a wide range of development pressures, a situation that exacerbates their vulnerability to climate change.

This section examines challenges and opportunities in reconciling development priorities with climate change adaptation in the case of coastal mangroves, using the Bangladesh Sunderbans and Viti Levu in Fiji as illustrative examples. The Sundarbans are situated in the delta of the Ganges, Brahmaputra and Meghna river system in the Bay of Bengal. They consist of a cluster of islands, tidal waterways and mudflats forming the largest continuous mangrove forest in the world, covering around 10 000 km^2. Roughly 60% of the Sundarbans are located in Bangladesh, with the rest lying in India. The Sundarbans support one of the richest natural gene pools of fauna and flora in the world: it includes numerous plant species and over 400 species of wildlife, including the Royal Bengal tiger. The area also offers subsistence livelihoods to about 3.5 million inhabitants. Viti Levu, an island in the tropical Pacific, is home to over two-thirds of Fiji's population. Some 230 km^2 of mangroves fringe its coastline. While its mangrove cover is smaller than that of the Sundarbans, it is quite significant for an island setting.

5.1. *Current climate-related threats*

Fiji and Bangladesh have tropical climates with relatively little inter-annual temperature variation. Both countries are subject to extreme climate events such as cyclones, floods and droughts. The location and topography of the Sundarbans make them particularly vulnerable to cyclones. These storm systems originate deep in the Indian Ocean and track through the Bay of Bengal, where shallow waters contribute to huge tidal surges when cyclones make landfall. Fiji, for its part, has the highest concentration of cyclones in the South Pacific, with major economic and public safety impacts (World Bank, 2000b).

The other extreme events influencing the hydrology of mangroves in both settings are droughts and floods. Both threaten the delicate balance between fresh and saline water necessary for mangroves to thrive. Bangladesh is located on a floodplain of three major rivers and their numerous tributaries. About one-fourth of the country is flooded every year. In extreme years, two-thirds of the country can be inundated. The exposure of Bangladesh to storm surges from the Bay of Bengal exacerbates this vulnerability. The low coastal topography also gives rise to a strong backwater effect, and seasonal variation in the interaction between brackish water and freshwater is considerable: freshwater dominates during the monsoon, and the saline front penetrates further inland during the dry season. In Fiji, droughts result from El Niño events, which typically cause hotter, drier conditions from December to February, and cooler, drier conditions from June to August. The 1997-98 El Niño resulted in one of the most severe droughts in Fiji's history.

5.2. *Implications of climate change*

Several dimensions of climate change will affect the geographic coverage and overall productivity of coastal mangroves. They include increases in air and ocean temperatures, sea level rise, changes in rainfall patterns, changes in salinity regimes (which would be influenced by changes in sea level and rainfall patterns, including inland), changes in cyclone frequency and intensity, and the impact of climate change on other elements of the coastal seascape, such as coral reefs, which could affect the viability of mangroves.

The rise in temperature is the most robust climate change projection. In principle, warming would make the climate of higher latitudes more amenable to mangroves. Any expansion towards temperate zones is unlikely to affect Bangladesh and Fiji, however, as they are situated in the tropics, where temperatures are already conducive to mangrove growth. Other

BRIDGE OVER TROUBLED WATERS – ISBN 92-64-01275-3 – © OECD 2005

possible impacts on mangroves resulting from higher temperatures, such as enhanced growth and productivity, are small in comparison to the impacts from sea level rise and altered hydrology.

Sea level in Fiji is projected to increase, with mid-range scenarios yielding predictions of 10 cm by 2025 and 50 cm by 2100, while scenarios based on higher GHG emission projections indicate a rise twice as high – over 20 cm by 2025 and 1 metre by 2100 (Feresi *et al.*, 1999). For Bangladesh, there is no specific regional scenario for net sea level rise, in part because the Ganges-Brahmaputra delta is still active and morphologically highly dynamic. The literature suggests that its coastal lands are receiving additional sediments because of tidal influence, while in some areas land is subsiding because of tectonic activities (Huq *et al.*, 1999). Since the delta is constituted by sediment decomposition, compaction of sediment may also play a role in defining net change in sea level for the coastal zone, including the Sundarbans. A review of the literature and of expert opinion suggests that the effects of sediment loading and of compaction and subsidence may cancel each other out, so that net sea level rise may be assumed. A US country study of Bangladesh put the range at 30-100 cm by 2100 while the IPCC Third Assessment gives a global average range with slightly lower values of 9-88 cm.

The likely response of mangroves to sea level rise depends upon a variety of factors – particularly the degree to which sea level rise is offset by increased sedimentation. If sedimentation rates exceed sea level rise rates, mangroves may even expand seaward. Mangroves in river-dominated settings such as the Sundarbans are more likely to keep pace with sea level rise because of the large volumes of sediment they receive from the rivers. For mangroves in small-island settings such as Viti Levu, sedimentation is more limited, and even where it keeps up with sea level rise, the latter is more likely to dominate in the long term. In the Sundarbans, it is projected that a 45 cm rise in sea level would inundate 75% of the mangroves, and a 67 cm rise could inundate the whole system. Extrapolating from this information, Smith *et al.* (1998) calculated that a 25 cm sea level rise would result in a 40% mangrove loss. Whether mangroves retreated shoreward would depend upon the rate of sea level rise and the presence of suitable land to retreat to. If the landward margins of mangroves are already at the boundary of steeply elevated, densely populated or highly developed land, inland migration will be severely constrained. Even if barriers to migration of mangroves, such as physical structures, could be moved, it is unlikely that inland migration would make up for losses of mangroves from inundation.

Closely intertwined with sea level rise is the issue of salinity levels in coastal mangroves. Some mangrove species may adjust to higher salinity levels in inundated regions, while others will not be able to. A further critical factor for the Sundarbans is that rising sea level would increase the backwater effect in the major Ganges tributaries, which would push the saline front further inland. This might be offset during the monsoon season, when, according to many climate models, higher rainfall and thus increased freshwater flows in the rivers are likely. The effects of climate change on the Sundarbans would be considerably greater during the dry season, for which climate models predict a decline in rainfall that would further reduce freshwater flows into the Sundarbans and make the saline front penetrate further inland. As a result, freshwater-loving woody tree species such as sundri are projected to decline and eventually disappear under climate change, to be replaced mainly by grasses, shrubs and trees of poor quality (Ahmed *et al.*, 1998). Since the composition of vegetation has profound effects on the distribution of forest fauna, such a change could seriously affect the long-term sustainability of the ecosystem.

Another dimension of climate change relates to changes in frequency and intensity of tropical cyclones. Bangladesh and the Sundarbans lie in an active cyclone corridor. Current climate models do not adequately resolve the question of the influence of climate change on cyclones (Risbey *et al.*, 2002). The historical record shows large inter-decadal variability, which makes any trend analysis based upon limited time-series data difficult to interpret conclusively. Still, basing its conclusion on emerging insights from some climate model experiments as well as the empirical record, the IPCC Third Assessment notes that "there is some evidence that regional frequencies of tropical cyclones may change but none that their locations will change. There is also evidence that the peak intensity may increase by 5% to 10% and precipitation rates may increase by 20% to 30%" (IPCC, 2001a). These estimates, however, are not location-specific. Any changes in frequency or intensity of cyclones, or both, are likely to compound salinity intrusion in coastal areas, including the Sundarbans and Viti Levu. It may be argued that the intensity of storm surges is likely to increase under climate change particularly in the later 21st century. Future cyclones and associated high-intensity storm surges would likely inundate high levees and back swamps that now do not get submerged with saltwater, further affecting salinity. Moreover, in areas where cyclonic storms cause widespread mangrove forest mortality, subsidence due to root decomposition will result in conversion to open water.

Finally, the health of mangroves will depend on how other seascape components respond to climate change. In this regard, the prime threat seems to be to coral reef systems. Rising sea surface temperatures have led to coral bleaching and mass mortality of reefs in some places. Some coral bleaching has occurred in Fiji (World Bank, 2000b). If this trend continues and Fiji's coral reefs die, it will have a profound impact on the seascape, since coral reefs help create sheltered areas where mangrove communities can establish themselves. Without this shelter, mangroves would be threatened by even more exposure and could end up dying as well.

5.3. *Mainstreaming adaptation responses in the Bangladesh Sundarbans*

The Bangladesh Sundarbans offer subsistence livelihoods for close to one million inhabitants. Traditional lifestyles of the people living in the fringe areas north of the forest boundary are well adapted to the tidal and seasonal variation in water and salinity levels. Dwellings are built on raised platforms, and farmers cultivate flood-tolerant rice during the monsoon on land protected by temporary dykes. Fishing of salt-tolerant varieties was long the principal source of livelihood during the dry season, when salinity levels were high.

These traditional lifestyles, however, are already under considerable pressure from non-climate trends including rapid population growth, poaching of wildlife and illegal felling of timber. In addition, industrial development in the region has increased demand for timber from the Sundarbans. Growing barge traffic, combined with limited environmental enforcement, has led to oil spills, which damage the ecosystem. Further pressure comes from shrimp farming, which took off as an export industry in the vicinity of the Sundarbans in the mid-1980s. While it raised incomes, it also encouraged deliberate inundation of land with brackish water during periods of low salinity to boost shrimp production. Yet another trend is water diversion upstream (including in India) from the rivers bringing freshwater to the Sundarbans, particularly during the dry season, when freshwater is needed to offset high salinity.

The impacts of climate change on the Sundarbans will be superimposed on these baseline stresses. Adaptation measures, therefore, need to be mainstreamed within a broader, system-wide response to the entire spectrum of climate and non-climate threats to the ecosystem.

The most useful adaptation strategy would be to reduce the threat of increasing salinity, particularly during low-flow periods. This might involve a range of physical adaptations to offset salinity ingress, including: *i)* increasing freshwater flows from upstream areas; *ii)* resuscitating existing river networks to improve flow regime along the forest; and *iii)* artificially enhancing these networks to facilitate freshwater flow regimes along the rivers supplying freshwater to the western parts of the forest. With regard to the first option, the Gorai River is particularly important. It is the only remaining major spill channel of the Ganges flowing through the region of the Sundarbans. Dry-season Gorai flows have been particularly affected by the building of the Farakka barrage in India. Following the signing of the Ganges Water Sharing Treaty with India in 1996, the flow of the Ganges within Bangladesh generally improved. Increasing Ganges flows from their current level will require increased co-operation among riparian countries (Ahmed, 2005). It will also require the creation of storage capacity in the Ganges Basin in Bangladesh so that a sustained flow regime can be maintained in the Gorai and other rivers throughout the dry season.

An adaptation aimed at saving the Sundarbans from submergence induced by sea level rise would be to modify the threat of permanent inundation. One possible approach is to enhance sedimentation on the floor of the mangrove forest through guided sedimentation. If this approach appeared technically feasible and economically viable at pilot level, projects could be undertaken in an effort to save the forest. Controlled and guided sedimentation would balance subsidence and could help delay permanent inundation.

Another way to enhance freshwater flows into the Sundarbans is to resuscitate existing river networks, which get clogged by heavy sediment from the Himalayas. The government and international donors have initiated engineering projects to this end, including the dredging of rivers such as the Gorai and the re-excavation of the Kobodak River, which can supply freshwater directly to the central Sundarbans. The Government of Bangladesh commissioned a study on options for "Ganges-dependent areas", which identified other possible ways to restore the Sundarbans. One proposal involves centralised freshwater pumping to maintain flow in the tributaries of the Ganges during the dry season (Halcrow and Associates, 2001). Another option under consideration is the construction of a Ganges barrage in Bangladesh to store water that could be released during the dry season. Meanwhile, the government is implementing an integrated coastal zone management plan that seeks to enhance livelihood opportunities while preserving the underlying natural resource base. The terms of reference for implementation require integration of adaptation to climate change. Still another approach is to protect the forest resources in the Sundarbans through species enrichment, increased surveillance and the establishment of a mangrove arboretum.

Some development policies and priorities could potentially conflict with the aim of reducing the vulnerability of the Sundarbans to climate change. In particular, plans to encourage ecotourism in the Sundarbans might risk adding another stress to an already fragile ecosystem.

While many structural adaptations, such as coastal embankments and salinity reduction, have been integrated in development projects and policies, their durability remains a challenge. For example, the high rate of sediment influx from Himalayan rivers means measures such as dredging of waterways must be repeated periodically. Monitoring and maintenance require continued government and donor interest, as well as participation of the local population, far beyond the duration of the original project.

The case of the Sundarbans also highlights the importance of the transboundary dimension in climate change adaptation. The effect of water diversion by the Farakka barrage on dry-season flows and salinity levels in the Sundarbans was comparable to, if not higher than, the impact projected to occur a few decades later as a result of climate change. Adaptation to climate change, therefore, may require not just local solutions but also cross-boundary

BRIDGE OVER TROUBLED WATERS – ISBN 92-64-01275-3 – © OECD 2005

institutional arrangements such as the Ganges Water Sharing Treaty to resolve the current problems of water diversion (Ahmed, 2002). Finally, addressing climate change risk should not overshadow other critical threats, from water pollution in surrounding areas to illegal logging and activities such as shrimp farming that have put the fragile ecosystem at risk even before significant climate change impacts manifest themselves.

5.4. *Mainstreaming adaptation responses for mangroves in Viti Levu*

Even as mangroves protect other coastal systems and human settlements from sea level rise and storm surges, they are themselves threatened by climate change through inundation, over-sedimentation, ecosystem breakdown and loss of coral reef cover. Hence, enhancing their resilience and protection is a good "no-regrets" adaptation, not only for the mangroves but also for other natural and socio-economic systems at risk from climate change. Adaptation to sea level rise and coastal impacts in Viti Levu would likely take two main forms: holding back the sea through physical barriers, or allowing the shoreline to retreat, which can involve the use of mangrove wetlands to provide a buffer against storm surges.

A detailed cost-benefit analysis of all adaptation options has not yet been carried out for Viti Levu, but studies conducted elsewhere generally indicate that allowing mangrove migration and a retreating shoreline would be more cost-effective than building physical barriers. In Fiji, there are two competing trends in this regard. At various levels of government and in local communities, a preference is emerging for managing sea level rise via mangrove conservation. Yet, the last century has seen destruction of mangrove regions and diminished mangrove coverage. The reasons are complex, but, in short, mangroves are losing out to pressures from agriculture, tourism and urbanisation.

The benefits of mangrove conservation tend to accrue either to small communities with little voice in government or to future generations with no present voice. The benefits of mangrove destruction tend to accrue to developers, companies and towns with more direct access to government and an ability to demonstrate tangible, immediate rewards from converting mangrove land. Thus, there are political barriers to conservation despite its being a no-regrets adaptation strategy. Also working against conservation is the fact that mangroves can be removed very rapidly, whereas it takes many years to regrow them.

The devaluation of mangrove services is illustrated by cost-benefit studies of these services in Fiji and a lack of policy coherence. Looking at whether mangroves should be preserved or converted for agricultural uses, Lal (1990) estimated that Fijian society would lose USD 181 000 per year if mangroves were clear-cut. Table 4.2 summarises three studies of the economic value of mangroves in Fiji. The results are broadly consistent. While none of the three captures all possible value from mangrove services, each includes the more salient mangrove services, associated with fisheries, habitat and coastal protection.

Table 4.2. **Estimated economic value per hectare of preserved mangroves for Viti Levu, Fiji**

Source	Economic value of goods and services	USD/ha/year
Lal (1990)	**Total economic value**	**2 706**
World Bank (2000b)	Subsistence fisheries	240-360
	Commercial fisheries	90-140
	Habitat functions	160
	Coastal protection	1 480
	Medicinal plants	230-330
	Raw materials	180
	Total economic value	**2 380-2 650**
Fiji Biodiversity Strategy Action Plan	Food, nutrient and habitat services	1 200
	Disturbance regulation (coastal protection)	1 250
	Total economic value	**2 450**

These estimates differ from the valuation used by the Department of Lands, which took over operational control of the mangroves in 1974 and is responsible for compensating villages when mangrove land is converted for other uses. The compensation is only for the loss of fishing rights, and it typically seems to amount to a one-time negotiated payment of around USD 150 000-200 000 for an area of about 70 ha of mangrove lost; that is, about USD 2 500/ha, a single payment rather than per hectare per year as in the Table 4.2 valuations. Assuming a 20-year replacement period for mangrove forest, this payment translates to USD 125/ha/year, about 1/20th of the annual value of mangrove services shown in Table 4.2. In effect, this undervaluation represents a subsidy for conversion of mangrove land. A key reason this subsidy has continued is a mismatch between the mangrove ecosystem and the system of property rights in Fiji. A traditional *mataqali* (clan) has communal claim over physical resources and the environment, including mangroves. However, the government has declared these rights as being *usus fructus* only, which affects the amount of compensation paid for the loss of mangroves to conversion (Lal, 1990).

In addition to inadequate valuation and the property rights regime, there is a lack of effective management and conservation. Considerable capacity is in place, however, and the issue is more capacity enhancement than development (Koshy and Philip, 2002). Moreover, active partnerships directed at mangrove conservation have been formed between villages and a sizable NGO community, backed up by interest from donors and capacity within government. Two pieces of recent legislation may also support mainstreaming: the Biodiversity Strategy Action Plan (BSAP), which has been approved, and the pending Sustainability Bill. If the valuation on which operational decisions about mangrove conversion are based were to be brought into line with the BSAP estimates shown in Table 4.2, it would provide a signal that mangrove conservation and coastal impacts are a high priority.

Given the competition between mangrove conservation and coastal development projects related to settlement, agriculture and tourism, it would be appropriate to situate mangrove management within a broader coastal management framework. Attempts to protect the coast are likely to prove more expensive over the long run than allowing the coast to recede; finding the right balance between development and conservation clearly involves a trade-off. Too little conservation of mangroves will lead to faster loss of coastal land and bigger impacts from storm surges.

Part of the solution, and hence of the planning vision, is to consider mangroves integral components of development projects. Mangroves will need to be able to migrate shoreward as sea level rises. If they encounter barriers such as settlements, roads and walls that impede migration, the rising sea will strand them. The protection and services they provide will then be lost. To provide mangroves with the ability to migrate, some setback needs to be declared that prohibits long-term development structures within the likely migration zones. The exact amount of setback required would depend on how much sea level rise is planned for. Nunn *et al.* (1993) note that in some countries, including Indonesia and Malaysia, coastal management plans call for mangrove buffers 50-100 metres wide, while current practice in Fiji is to maintain a belt 5-30 metres wide.

Finally, a coastal management plan is needed that prioritises mangrove conservation, requires sufficient no-construction zones above the high-water line to facilitate mangrove migration, and engages local communities in both processes. At the local level, Fijian villages dominate the coastal environment. Some work together and with NGOs to conserve marine and coastal resources. Although the number of projects of this kind is relatively small at this point, those that exist have been widely viewed as successful. Concerns have been expressed that some of these efforts may have focused too narrowly on restoration of mangrove species, to the disadvantage of the habitat that supports them. If the broader ecosystem is not considered, local efforts at mangrove conservation may be hindered by non-local stresses, such as upland erosion, silting and agricultural run-off, that cause habitat deterioration. Such concerns are being recognised, and opportunities exist to promote more initiatives, as a way both to extend the area of mangroves under local management and protection and to raise the profile of mangrove conservation with government.

6. Mainstreaming GHG mitigation in agriculture and forestry in Uruguay

The cases described in this chapter so far have focused on potential synergies and conflicts between natural resource management policies and responses that might be required for *adaption* to climate change. The next case examines instead the links between natural resource management and *mitigation* of GHG emissions in the agriculture and forestry sectors of Uruguay.

Uruguay is situated in south-eastern South America. It is bordered by Argentina to the west, Brazil to the north-east, the Río de la Plata to the south and the Atlantic Ocean to the east. Many of its demographic characteristics – such as the annual population growth rate, life expectancy, infant mortality and GDP per capita – are closer to those of OECD countries than to those in the developing world. Like many developing countries, however, Uruguay bases much of its economy on natural resources, particularly on agriculture and livestock, and forestry.

The potential impacts of climate change on these sectors in Uruguay are uncertain. Some principal crops are projected to be negatively affected, but there is low confidence in these projections (Baethgen, 1997). At the same time, agriculture and forestry offer considerable potential for GHG mitigation. Agriculture alone accounts for close to 80% of total GHG emissions in Uruguay. Agricultural and forestry policies have had a remarkable ancillary impact on carbon sequestration. The following sections examine these two sectors, with particular focus on the implications of current policies on carbon sequestration. They also assess opportunities to further promote GHG mitigation while responding to economic development and conservation priorities.

6.1. *Overview of the agriculture and forestry sectors*

Uruguay's economy is largely dependent, directly or indirectly, on crop and livestock production. Some 85% of the land is suitable for agricultural production – one of the largest proportions in the world. The highly fertile soil of the pampas provides an environment in which native temperate and subtropical grasslands are used for livestock grazing or have been converted to improved pasture and to crop land. The main crops include wheat and barley in winter and rice, maize, sorghum, sunflower and soybeans in summer. Livestock rearing dominates the sector, accounting 90% of the agricultural land and generating 60% of total agricultural value and 70% of agricultural exports. Even in the areas with deep, fertile soil where annual crops are grown, rotation between crops and pasture is practiced, currently three to four years of cereal and oil crops alternating with three to four years of sown pasture (mixed grass and legumes) used for relatively intensive beef production. Taken together, crop and livestock production satisfy almost the entire domestic demand for food and support an agro-industrial sector that generates about 60% of the total industrial product.

The predominant natural ecosystem in Uruguay is grassland, which covers about 85% of the country. Natural forests occupy 3.5%, about 6 700 km^2. They include riparian forests, subtropical ravine forests in the north, highland forests in the south and palms in the east and west. Tree planting was introduced in the late 19th century. Small areas of eucalyptus were established on ranches to provide shade and shelter for cattle and wood for fences and cooking fires. At the same time, pine plantations and, to a lesser extent, eucalypts were established in southern coastal areas to stabilise sand dunes. The ranch and dune plantations totalled 900 km^2. Commercial forestry with large plantations started in the mid-20th century. The first investors included pension funds, small pulp mills, other private investors and the national utility company. In 1967, the first regulation of commercial forest plantations was enacted. It provided incentives to invest in plantations through a partial income tax exemption, proportional to the annually planted area. This resulted in a doubling of the annual planting rate, to 27.5 km^2, from 1968 to 1979, after which the incentive was withdrawn. By 1988, commercial forest covered about 310 km^2, fairly evenly distributed across the country. Most of this area was planted with short rotation eucalyptus (10-12 years) and pine (25-30 years). The low quality of tree varieties planted, intensive tillage and damage from cattle grazing contributed to relatively low growth rates and poor quality of harvested timber.

6.2. Implications of sectoral policies on carbon sequestration

Several policies concerning agriculture, livestock and forestry over the past three decades have noticeably affected carbon sequestration in Uruguay. Today, land-based carbon sequestration (in grassland, agricultural soil and forest) amounts to about 14-15 million tonnes of CO_2 per year, or about 2.5 times the total annual CO_2 emissions. In terms of all GHGs, carbon sequestration offsets around half of current emissions, measured in global warming potential (Baethgen and Martino, 2004). The following sections outline some key policy initiatives that have contributed to growth in carbon sequestration.

6.2.1. Agriculture and livestock sector

Early government interventions in the agriculture sector were oriented towards improving natural grassland, the backbone of livestock production. The most common instruments were tax exemptions and low-interest loans for the introduction of legumes, grasses and phosphate fertiliser. These measures had a significant impact: the areas of improved pasture doubled between the 1950s and the late 1970s. Increasing the area under sown pasture and improving much of the natural grassland (by introducing legumes and phosphorous fertiliser) raised the soil nitrogen content, thereby increasing carbon levels in previously nitrogen-limited soil. However, as domestic prices for beef and other livestock products were low, economic incentives for farmers to invest in technological improvements or improve livestock productivity were nonexistent. In terms of GHG emissions, these early initiatives were characterised by: *i)* some increase in the amount of carbon sequestered in soil, mainly due to the rise in the nitrogen content of grassland soil; and *ii)* little if any change in methane emissions from livestock, since beef and wool productivity stagnated between the 1960s and early 1980s.

A combination of domestic price rises for livestock products and the opening of regional and international markets in the late 1980s and 1990s provided economic incentives for rapid growth in the area of improved pasture and the efficiency of livestock production. The results included decreased numbers and increased health of herds and the use of improved feed. Total milk production, for example, increased by 300% over 1985-2001, but methane emissions per litre of milk decreased by 15% over 1990-2000. Similarly, beef production grew by 248% from 1960 to 2000, but methane emissions per kilogram of beef decreased by 10% over 1990-2000.

Meanwhile, the 1982 Soil Conservation Law allowed the Republic Bank to require the use of soil conservation techniques as a condition in its rural credit programme. The resulting widespread application of soil management techniques reversed a trend of soil degradation and dramatically improved soil carbon levels. For example, growth in the annual crop area with reduced or no tillage caused soil carbon sequestration to rise by 200-600 kg of carbon per hectare per year (the range represents variation in soil textures and previous land use). Overall, as a result of improvements to pasture and the use of conservation techniques, the estimated amount of carbon sequestered in soil increased by an average of 1.5 million tonnes per year over 1966-2000 (Figure 4.4).

6.2.2. Forestry sector

A major breakthrough in the forestry sector was the adoption in 1987 of a forestry promotion policy based on instruments contained in the newly passed Law No. 15939 (Box 4.3). The central objectives of the policy were to provide raw material for export products and provide a sustainable supply of firewood while stopping deforestation, which had been increasing. The policy was highly successful and resulted in remarkable growth in the forested area (Figure 4.5). The estimated investment involved, including a significant amount from foreign sources, totalled more than USD 1 billion in the 1990s. New technology-based practices resulted in better quality and in more vigorous and more homogeneous tree stands. Productivity rose by as much

Figure 4.4. **Changes in soil carbon content due to land use change**

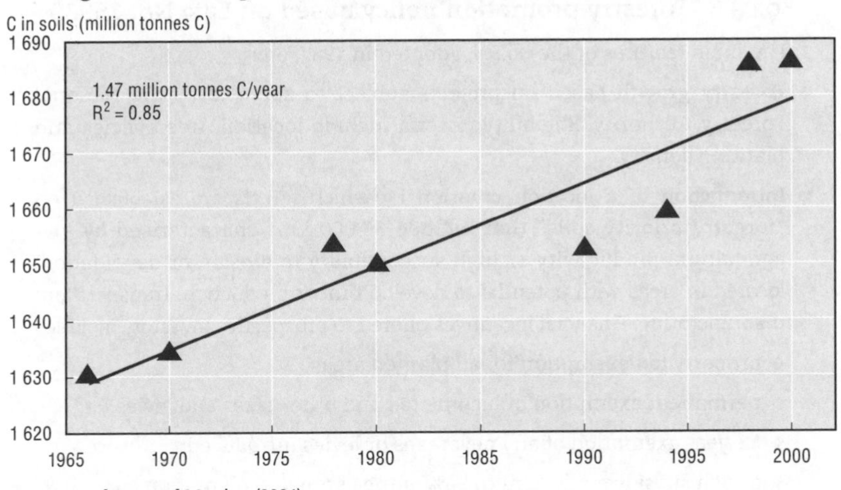

Source: Baethgen and Martino (2004).

as 100%. Forest companies introduced modern concepts such as long-term planning, environmental management systems and attention to working conditions and other social issues.

The forest plantation area grew dramatically after 1990. This growth, combined with high tree biomass productivity and relatively low initial soil carbon, resulted in considerable further carbon sequestration. A study by the Ministry of Environment estimated that cumulative net carbon sequestration by forestry over 1988-2000 amounted to 7.4 million tonnes (27.4 million tonnes of CO_2). The study projected additional net sequestration of 29.3 million tonnes of carbon (108.6 million tonnes of CO_2) for 2001-12 (Uruguay, 2002).

This vigorous development of plantation forestry has climate change implications beyond carbon sequestration. Residue from forest harvesting and wood manufacturing can be used for heat and power generation and, eventually, for biofuel production, offsetting GHG emissions from fossil fuel burning. The increased availability of better-quality wood products, combined with development of wood manufacturing capacity, may result in more use of wood in construction instead of the much more energy-intensive bricks and cement. This would extend the residence time of carbon in wood products, and the higher thermal insulation of wood would mean energy efficiency improvement as well. Also, local use of locally produced wood products would reduce GHG emissions from transport.

Box 4.3. **Forestry promotion policy based on Law No. 15939**

The main features of the policy, adopted in 1987, are:

- **Forestry growth** based on projects subject to approval by the National Forestry Authority. Eligibility criteria include location, tree species and planting density.
- **Introduction of a location criterion** by which forests are established on "forestry priority soils" that include 36 000 km^2 characterised by low agricultural productivity or high susceptibility to erosion or degradation, located in areas with potential to develop timber production, transport and manufacturing. Financial incentives offered to prospective investors include:
  - property tax exemption for all planted areas;
  - permanent exemption of income tax and other taxes and levies;
  - 12-year exemption of any new taxes or levies introduced;
  - a cash subsidy equivalent to 50% of the estimated plantation cost;
  - duty-free import of goods to be used in approved projects;
  - soft credit for planting, with a grace period of 10 years on both principal and interest;
  - permission for corporations to buy land if forestry is their main activity;
  - separation of forest ownership from land ownership to provide flexibility for using financial mechanisms. A later regulation allowed rental contracts for up to 30 years for forestry (for other purposes, the maximum is 15 years);
  - permission for investors to deduct up to 30% of their income tax payments on other activities for investments in forestry projects. A similar benefit is provided to buyers of Uruguay's external debt bonds.
- **Prohibition of harvesting in native forests** except for wood supply for farms and certain cases subject to Forestry Bureau approval.
- Enforcement of **fire and pest prevention** measures.
- **Promotion of the climate benefits** of forests as recognised in the introductory message of the Executive to Parliament and in Article 4 of the Law. A study by JICA (1991) estimated that planting 1 000 km^2 over 1991-95 would offset half of the CO_2 emissions from fossil fuel combustion in Uruguay.

Figure 4.5. **Evolution of commercial forest plantation area in Uruguay over 1975-2002 and projected "business-as-usual" plantation to 2010**

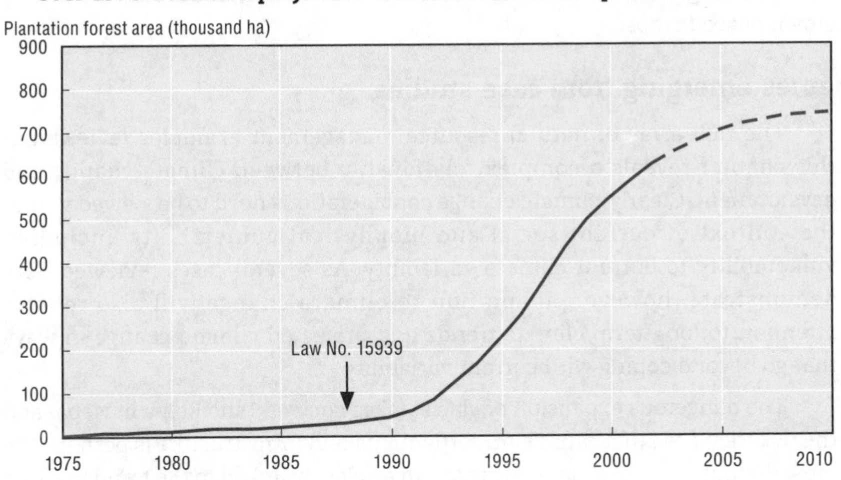

Source: Baethgen and Martino (2004).

6.3. Outlook

Sectoral policies for agriculture and forestry have resulted in carbon sequestration equivalent to about 2.5 times the total annual CO_2 emissions in Uruguay. The increase in the area of improved pasture, the decrease in the area of annual crops and the use of better tillage practices have resulted in increases in soil carbon sequestration that help offset CO_2 emissions from energy use and industrial processes. In addition, the dramatic growth in the forest plantation area from 1990, combined with high tree biomass productivity and relatively low initial soil carbon content, resulted in further removal of atmospheric CO_2. The Ministry of Environment estimated cumulative net carbon sequestration by forestry over 1988-2000 at 27.4 million tonnes of CO_2 and projected additional net sequestration of 108.6 million tonnes in 2001-12.

There is further potential to sequester carbon while continuing to meet national development and conservation priorities. Better silvicultural management practices, coupled with financial incentives such as availability of long-term loans, have the potential to significantly improve forest productivity and carbon sequestration rates. Government policies could facilitate substitution of sustainable wood for energy-intensive products such as cement in construction. Further increases in the area of improved pasture and better soil conservation could produce more growth in carbon sequestration in soil, while better feed and younger herds could translate into reduced methane emissions from livestock. There is also considerable potential for GHG mitigation by promoting the use of wood and rice husks in electricity generation and replacing conventional diesel fuel with bio-diesel produced from locally grown oilseed crops.

7. Themes emerging from case studies

The rich array of natural resource management examples reviewed in this chapter reveals a complex relationship between climate change and development. Clearly, climate change considerations need to be viewed within the context of current social and biophysical vulnerability, including vulnerability to current climate variability. As several cases reviewed here demonstrate, however, adaptation to climate change will also require attention to long-term climate trends and projected climate change in ways that go beyond coping with current variability.

The dangerous expansion of glacial lakes such as Tsho Rolpa in Nepal and the risk they pose to downstream settlements and infrastructure is perhaps the most dramatic example, and one that can be closely linked to the trend in rising temperatures. The long-term implications of such trends in "creeping hazards" clearly need to be factored into the design of infrastructure projects and development choices in addition to being taken into account when coping with climate variability. Similarly, on Mount Kilimanjaro, adaptation responses need to be implemented now to offset the steady loss of subalpine cloud forest to forest fires due in part to the long-term trend towards a warmer and drier climate. In other cases, countries are reasonably well adapted to the impacts of inter-annual climate variability, such as in Egypt, where Nile water availability has been controlled through the construction of the Aswan High Dam. Nevertheless, there is a critical need to consider the implications of any long-term reductions in Nile water availability in Egypt under climate change. Such reductions, stemming from increases in water demand and evaporation losses, along with (possibly, though less certainly) changes in rainfall, would only compound water stress resulting from demographic pressures and the prospect of rising water abstraction in upstream riparian countries. In other words, adaptation strategies to consider medium- and long-term implications are important even where climate change itself may be only the proverbial "straw that breaks the camel's back".

The cases examined in this chapter also highlight the fact that the complexity of the relationship between climate change and development extends both to the level of livelihoods linked directly to natural resources and to the macro level in terms of infrastructure and sectoral policies. The cases of Mount Kilimanjaro and the Sundarbans illustrate the role that climate change impacts could have in the progressive degeneration of forest cover and hence of the ecosystem services the forest provides. On Kilimanjaro, loss of the cloud

BRIDGE OVER TROUBLED WATERS – ISBN 92-64-01275-3 – © OECD 2005

forest would have a major effect on water resources; in the Sundarbans, replacement of woody species by poor-quality forests would affect wood availability, the sustainability of the ecosystem and the livelihoods that depend upon it. By the same token, livelihood choices and development policies play a critical role in determining the vulnerability of natural resources to climate change. Activities such as upstream water diversion have led to the same degree of vulnerability in the Sundarbans today that climate change could have been expected to produce in several decades. Even if warmer, drier conditions on Kilimanjaro are more conducive to the spread of forest fires, the fact remains that people start many of the fires.

Another related issue is the degree of policy coherence between development priorities and climate change objectives. Perhaps the most remarkable example here is that of the agriculture and forestry sectors of Uruguay, where soil conservation, zero tillage and plantation forestry policies, initiated largely for sectoral and development purposes, led to a dramatic rise in carbon sequestration. Also noteworthy is the explicit recognition of "climate benefits" from forest plantation in a law passed a decade before ideas like the Clean Development Mechanism gained international currency. Elsewhere, initiatives synergistic with adaptation to climate change include efforts to boost dry season river flows and thus reduce salinity levels in the Bangladesh Sundarbans, reduce glacial lake hazards in Nepal, plant mangroves in Fiji, promote water conservation in Egypt and increase forest conservation on Kilimanjaro.

Examples of a lack of policy coherence between the two sets of objectives also exist, however. In Fiji, growing attention to the impacts of climate change is at odds with the significant undervaluation of coastal mangroves in operational decisions, contributing to the loss of a "no regrets" option to protect the coast from sea level rise. In the Sundarbans, policies to promote ecotourism may add one more stress to an already fragile ecosystem that is also vulnerable to climate change. In Nepal, adaptation to the greater variability of river flows projected under climate change might require consideration of storage hydropower or dams that could conflict with other development and environmental priorities. It is important to recognise that, while adaptation and development can indeed be synergistic, in some instances they may also conflict.

Finally, several cases examined in this chapter underscore the importance of the transboundary dimension of mainstreaming, which so far has been largely overlooked. The case of Egypt highlights the critical importance of transboundary water sharing arrangements. The vulnerability of the Sundarbans is linked to upstream water use. Many glacial lake outburst floods in Nepal originated in Tibet. These examples demonstrate the urgency of emphasising mainstreaming throughout a hierarchy of levels that is not limited to local and national but also includes a regional dimension.

Glacial Lake Outburst Floods in Nepal

Plate 1: Breaching of moraine dam, Tam Pokhari glacial lake (3 September 1998);
Plate 2: Houses being washed away by the Tam Pokhari glacial lake outburst flood;
Plate 3: Tam Pokhari after the breach.

Reducing Glacial Lake Outburst Flood hazards in Nepal

Plate 4: The Tsho Rolpa glacial lake, at an altitude of 4 850 meters and containing close to 100 million cubic meters of water, is one of the most dangerous in Nepal;
Plate 5: Drainage to lower the level of the Tsho Rolpa as an anticipatory adaptation to reduce the risk of glacial lake outburst flood.

Mount Kilimanjaro, Tanzania

Plate 6: Savanna grassland on the foothills of the snow-capped mountain: tropical heat and arctic conditions are only 30 km apart; *Plate 7*: The shrinking ice cap.

Forest fires on Mount Kilimanjaro

Plate 8: Burning subalpine Erica forests and bush on the north eastern slope;
Plate 9: Burnt Erica forest one year after a fire.

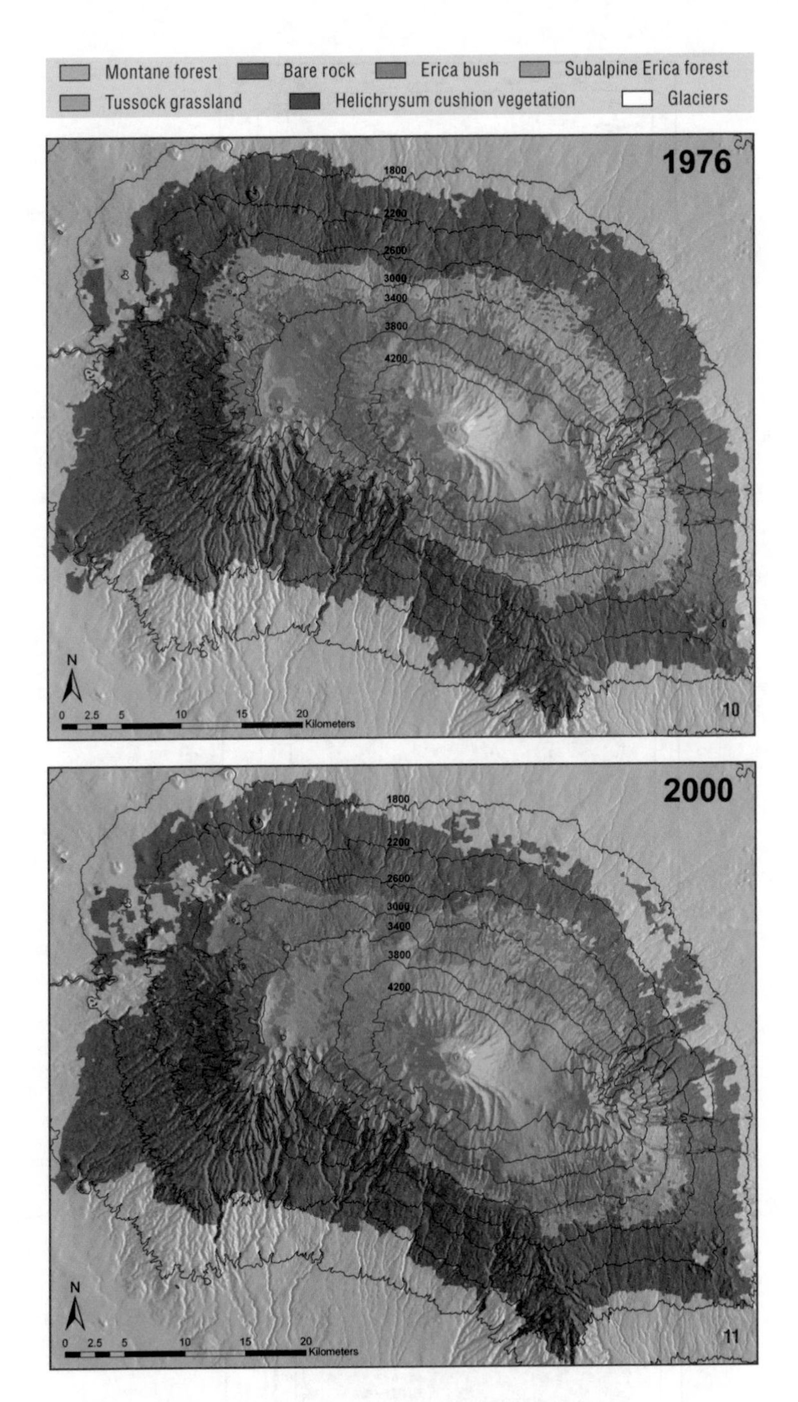

Vegetation cover on Mount Kilimanjaro

Vegetation cover in 1976 (*Plate 10*) and 2000 (*Plate 11*) showing loss of the Erica forest belt.

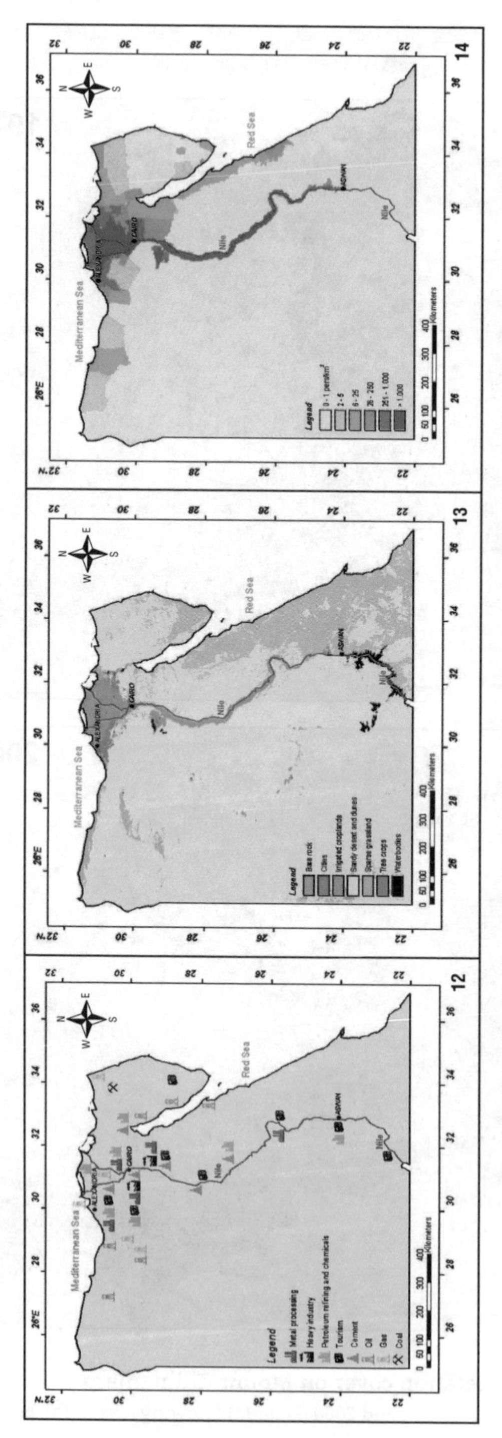

Spatial distribution of economic activity, land use and population in Egypt

Economic activity (*Plate 12*), land use (*Plate 13*) and population (*Plate 14*) in Egypt are concentrated along the Nile and the Nile delta which are also particularly vulnerable to climate change.

Community level adaptation in Fiji

Plate 15: Mangrove plantation in Viti Levu;
Plate 16: Community members in Kabara formulating an action plan to increase their resilience to the impact of climate change.

Greenhouse gas mitigation in forestry and agriculture sectors in Uruguay

Plate 17: Eucalyptus forest plantation with grazed interrows; *Plate 18*: Greenhouse gas emissions from livestock have been reduced through better feed and quality of livestock. Here a Holstein cow is shown with a device to measure methane emissions.

BRIDGE OVER TROUBLED WATERS – ISBN 92-64-01275-3 – © OECD 2005

ISBN 92-64-01275-3
Bridge Over Troubled Waters
Linking Climate Change and Development

Chapter 5

Bridging the Gap Between Climate Change and Development

by
Shardul Agrawala and Maarten van Aalst

This volume has explored the synergies and trade-offs involved in mainstreaming climate change in development activities, focusing on natural resource management. Findings from this work underscore the need for, and the challenges faced in, taking climate change into account in development activities. This concluding chapter summarises the findings and then outlines some of the principal barriers facing the mainstreaming of climate change adaptation in development activities. The chapter concludes with an agenda for further action organised around improving the usability of climate information, developing and testing climate risk screening tools, employing appropriate entry points for climate information in development activities, focusing more on implementation, and improving co-ordination and sharing of good practices.

1. A summary assessment

Climate change is inextricably linked with development choices and pathways. Decisions about energy use, transport infrastructure or forestry, for example, can critically affect GHG emissions, and hence the rate and magnitude of climate change. Climate change in turn will have additional impacts on natural and socio-economic systems that are already subject to a range of other stresses, including climate variability. Addressing the problem will require both the yin and the yang of climate policy – *mitigation* to reduce the build-up of GHG in the atmosphere, as well as *adaptation* to the inevitable impacts of climate change. Integrating or "mainstreaming" such measures into regular development activities can make the activities more robust and amplify their eventual impact. It can help pull in a range of new actors from non-environment departments within governments as well as the private sector and civil society who might otherwise not be engaged in the climate issue. Mainstreaming also offers the opportunity of greater policy coherence between development and climate objectives.

This volume has explored the synergies and trade-offs involved in mainstreaming climate change in development planning and assistance, with natural resource management as an overarching theme. The focus was primarily on mainstreaming adaptation, although links between development objectives, natural resource management and GHG mitigation were also considered. Case studies in Bangladesh, Egypt, Fiji, Nepal, Tanzania and Uruguay examined several dimensions of the climate change-development link, including regional climate change scenarios and impacts; attention to climate risk in national and development co-operation plans and projects; and in-depth analyses of the mainstreaming challenge in the context of critical ecosystems and sectors.

Several findings that have emerged from this body of work reinforce both the need for, and the challenges faced in, taking climate change into account in development planning and activities. An analysis of the composition of ODA flows to the case study countries indicates that a significant portion is in sectors potentially affected by climate risks, including climate change. While any classification based on sectors suffers from oversimplification, it is evident that consideration of climate risks would often be important for development investments and projects.

Another finding from this volume is that, in addition to short-term variability in climate, long-term trends and climate change are having a discernible impact on development activities. This is particularly the case for Nepal, where glacier retreat and increased risk of glacial lake outburst flooding have been observed, concomitant with significant temperature increase in the middle and high Himalayas. Clearly, a diverse range of development activities, from national planning to the design of hydropower systems to programmes concerning rural development, may need to adapt to the implications of both current and future climate risks.

Even in cases where the impacts of climate change itself are not yet discernible, scenarios for future impacts may already be sufficient to justify building some responses into planning. One reason is that it could be more cost-effective to implement many adaptation measures early on, particularly for long-lived infrastructure. Another reason is that, in many contexts, current development activities and pathways might irreversibly constrain future adaptation to the impacts of climate change. This could be true, for example, in the case of destruction of coral reefs or coastal mangroves, or development of infrastructure and settlements in areas that seem likely to be particularly exposed to the expected impacts of climate change. In such instances ameliorative measures might be needed in the near term, keeping in view the long term implications of climate change.

Do development activities pay sufficient attention to climate risks and, in particular, climate change? In a way, societies have always factored some climate information into their plans and practices. Weather and climate considerations influence clothing and housing choices. They are also typically part of the information used when farmers select crops and farming practices, engineers design highways and energy utilities determine generation capacity and demand. In other words, to a large extent, *some* climate considerations are routinely taken into account in day-to-day decisions. On the other hand, there is strong evidence that not all climate risks are being incorporated in decision making, even with regard to natural weather extremes. Nor, it would appear, are practices that take into account historical climate necessarily being suitably adapted to changes in these patterns resulting from climate change. Many planning decisions by governments, communities, businesses and individuals tend to focus, moreover, on shorter timescales and tend to neglect the longer-term perspective.

An assessment of a range of development activities in the six case study countries reveals a fairly nuanced picture in terms of attention to climate change concerns. Considerable progress has been made over the past decade or so with regard to activities specific to climate change, including assessment of mitigation measures as well as of climate change impacts and adaptation. Institutional mechanisms to address climate change have been established, and development of climate change action plans, and even more specific National Adaptation Programmes of Action (NAPAs), is under way in some cases. Development co-operation agencies have been active partners in many of these activities through their climate change programmes. These initiatives, however, are often confined to the climate change community, with relatively little outreach to sectoral decision makers. Adaptation measures also remain largely theoretical at this stage: progress in actual implementation has been quite limited.

From the development side, an analysis of national development plans, Poverty Reduction Strategy Papers, sectoral development strategies, country assistance strategies and project design documents in vulnerable sectors indicates that such documents pay negligible attention to climate change and often only limited attention to current climate risk. Some exceptions include the Interim-PRSP of Bangladesh, its 1999 national water policy and a subsequent national water management plan, which recognise direct links between climate change and key development priorities. However, specific operational guidance on incorporating climate change considerations into poverty alleviation strategies or water management decisions is generally lacking. Similarly, development co-operation agencies are increasingly recognising climate change as a serious challenge for their core activities but are only beginning to address the question of how to actually carry out mainstreaming.

2. The challenge of implementation

Why is it so difficult to implement and mainstream responses to climate change – particularly adaptation – within development activity? Lack of awareness of climate change within the development community and limited resources to implement response measures are the most frequently cited explanations. They may well hold true in many situations, but underlying them is a more complex web of reasons.

2.1. Relevance of climate information for development-related decisions

Development activities can be sensitive to a broad set of variables related to climate, from temperature and rainfall to sea level rise, snow cover, sea ice, stream flow and wind intensity. In climate models, however, certain variables can be projected better than others. For example, temperature is generally easier to project than sea level or rainfall, which in turn might be easier to project than variables like wind intensity. Mainstreaming of climate considerations may therefore be more difficult where the climate sensitivity of development-related decisions is to variables that cannot be reliably projected.

Often development activities are more sensitive to changes in climate extremes than to trends in average climate conditions. Agriculture, for example, may be more sensitive to the risk of extended hot and dry spells than to changes in seasonal averages of temperatures and rainfall. Deviations or changes in extremes, however, are often more difficult to predict through climate models than mean trends, though progress is being made on improving the quality of projections of certain extremes, particularly regarding temperature.

In addition, there is often a mismatch between the timescales of climate change projections and the planning horizons for many development activities. Development decisions are generally oriented towards the near and medium term, while many climate change impacts will become more significant only over the medium to long term. There are exceptions, of course, in both cases. For example, many climate change impacts, such as those related to glacier retreat and expansion of glacial lakes, are not decades into the future but pose a threat to development today. Conversely, not all development activities have short- to medium-term horizons. Infrastructure decisions, for example, have implications over several decades or even a century or more. Nevertheless, the general point remains that the timescales for development-related decisions and those over which climate change impacts are expected to manifest themselves may be incongruent.

2.2. Uncertainty of climate information

Closely related to the relevance of climate information is the issue of the uncertainty associated with projections of climate change. The authoritative assessments of the IPCC have regularly documented significant advances in the detection and attribution of observed climate changes at global level, as well as in the projection of future climate changes and their impacts on different sectors and world regions. This information has played a key role in highlighting the need for mitigation through multilateral and domestic actions. However, sensitising decision makers to climate change and its implications for variables such as global mean temperature is typically not sufficient to influence adaptation actions or to alter development practices on the ground. Decision makers generally need more precision on the implications of climate change (and associated uncertainties) for the locations, time frames and spatial scales that directly concern them.

A key point here is that climate change projections have widely differing degrees of uncertainty associated with them. The extent of uncertainty can depend upon the spatial scale of the projection, the region and/or season for which a projection is made and the variable being projected. Large-scale climate change projections typically have lower uncertainty than those specific to a particular location. In contrast, few development activities are directly sensitive to global or even continental-scale climate averages. Rather, the primary sensitivity is at a more local scale (such as a watershed or a city), for which credible climate change projections are often lacking. Furthermore, certain regions may have higher predictability than others. Finally, as has been noted, uncertainties also depend upon the variable being projected. In general, projections of temperature tend to have lower uncertainty than those for precipitation.

Nuances like these are often insufficiently understood or inadequately conveyed by those communicating climate information to the end users who might be expected to make decisions based on it. This lack has already been observed in the context of application of forecasts of seasonal climate and natural variability (Broad and Agrawala, 2000), and it is becoming increasingly evident in the context of climate change.

There is often an "uncertainty trough" (Figure 5.1), with climate change communicators ascribing lower uncertainty to a projection than the climate modellers themselves. This occurs in part because the communicators may not fully appreciate the complexity of the information they are transmitting, and perhaps in part because they may be driven by a supply push. Meanwhile, end users of such information who face real stakes may tend to be much more cautious in their acceptance of the information and implicitly ascribe much higher uncertainty to it. Both sides of the uncertainty trough can limit the uptake of climate information in adaptation decisions, even in cases where decision makers may otherwise be well sensitised to the overall seriousness of the problem.

Figure 5.1. **Uncertainty perception among producers, communicators and users of climate information**

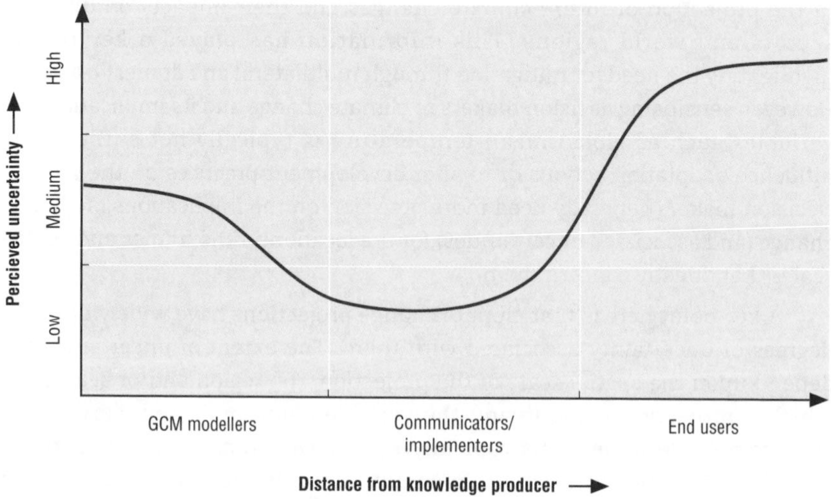

Source: Agrawala (2004), after Shackley and Wynne (1995).

2.3. Compartmentalisation within governments

Climate change is often viewed primarily as a top-down, multilateral-negotiations issue. Environment ministries have generally served as the focal points of the UNFCCC process (Table 5.1). They typically have limited influence with the more powerful "line ministries", such as those dealing with finance, transport and agriculture, whose policies and regulatory frameworks might need to be modified for successful integration of climate change considerations. Having multilateral negotiations as the prime driver for climate policy could have particular implications for the integration of adaptation considerations that are more bottom-up and do not require global co-ordination. Furthermore, the complex institutional structure and polarisation of international climate change negotiations can delay implementation even of "no regrets" adaptation options. The reason is that governments may wait for multilateral action on decisions under the climate regime instead of going ahead with measures that would make good economic sense to achieve other development objectives.

Table 5.1. **Location of UNFCCC national focal points within governments**

Environment agency or ministry	72%
Foreign affairs ministry or embassy	14%
Meteorological service	8%
Other/unknown	6%

Note: Based on UNFCCC (2005).

2.4. Segmentation and other barriers within development co-operation agencies

International donors increasingly seek to respond to needs and priorities expressed by recipient countries themselves, preferably through sectoral strategies and PRSPs. If climate change concerns do not feature prominently among national priorities, development co-operation agencies tend not to put them high on their own agendas.

Development co-operation agencies do set standards for projects and programmes and could make climate change considerations part of the operational requirements. Sectoral segmentation within the agencies, however, may limit their ability to do so. Climate change is often dealt with by one or two specialists who may not have direct leverage over operational guidelines and projects within their own agencies. Sectoral managers and country representatives may also face "mainstreaming overload", with competing agendas such as gender and governance also vying for integration within core development activities. Some observers suggest that documents such as PRSPs are already overburdened, and that adding more issues would undermine their credibility and effectiveness.

Another challenge relates to the structure of development co-operation itself. Although there is a move towards programme assistance, much aid still flows through projects lasting three to five years. While it would certainly help to incorporate climate risks (where appropriate) in such projects, they may not be the best type of vehicle for long-term strategic risk reduction. Funding priorities are a further issue. Anticipatory adaptation to climate change, like other forms of risk reduction, might have more difficulty attracting resources than more visible investments in emergency response, post-disaster recovery and reconstruction, where funding modalities are better established.

2.5. *Trade-offs between climate and development objectives*

Mainstreaming could also prove difficult to carry out because of direct trade-offs in certain cases between development priorities and the action required to deal with climate change. Governments and development co-operation agencies confronting pressing challenges such as poverty, lack of basic services and inadequate infrastructure may have few incentives to divert scarce resources to investments that are perceived as not paying off until climate change impacts fully manifest themselves. Putting a real value on natural resources and deciding when *not* to develop coastal areas or hillsides may also be seen as hampering development. At the project level, mainstreaming might be thought of as complicating operating procedures with additional requirements or considerations, or raising costs. In addition, short-term economic benefits that often accrue to a few in the community can crowd out longer-term considerations such as climate change. Activities like shrimp farming, mangrove conversion and infrastructure development, for example, may provide employment and boost incomes, but they also promote maladaptation and increase the vulnerability of critical coastal systems to climate change impacts.

3. A time of opportunity

Despite the wide-ranging challenges to successful mainstreaming of adaptation and other climate change considerations in development activities, some recent developments give grounds for optimism. Concern about climate change is growing within the development community. The 2003 report by ten bilateral and multilateral development and environment agencies led by the World Bank affirmed the central importance of climate change impacts and adaptation to the achievement of development goals for poverty alleviation (Sperling, 2003). It led to the establishment of the Vulnerability and Adaptation Resources Group, in which donors meet regularly to share their experiences with integrating adaptation into their core activities. In 2004, the Council of the European Union reiterated that "mainstreaming of responses to climate change into poverty reduction strategies and/or national strategies for sustainable

Box 5.1. **EU action plan on climate change in the context of development co-operation**

In November 2004, the General Affairs and External Relations Council adopted an action plan for the EU strategy on climate change in the context of development co-operation for 2004-08. The plan translates recommendations of the strategy document into concrete actions. It focuses on adaptation to climate change, capacity development and research.

The strategic objectives of the plan include: *i)* raising the policy profile of climate change and examining "synergies at the implementation level between various development related actions under the different multilateral environmental agreements and other international initiatives"; *ii)* supporting adaptation to climate change, with one goal being to develop guidelines for integrating climate change into development programmes; *iii)* supporting mitigation and low-GHG development paths, including the elaboration of guidelines to facilitate mainstreaming of low-GHG development into country strategy papers and national indicative programmes; and *iv)* developing capacity.

The Council of the European Union agreed that "EU member States and the Commission shall collectively implement the Action Plan in a co-ordinated and complementary manner and in line with their respective development co-operation programmes and priorities".

Source: Council of the European Union (2004).

development is the main avenue to address both adaptation to the adverse effects and mitigation of the causes of climate change" (Council of the European Union, 2004). An EU action plan was subsequently developed to implement the EU strategy on climate change in the context of development co-operation (Box 5.1).

Development agencies, in particular the World Bank, are developing and applying climate risk screening tools for their activities (Burton and van Aalst, 2004). In a related context, the momentum for disaster risk reduction more generally has been building since the 1994 Yokohama conference for the UN International Decade on Natural Disaster Risk Reduction, which culminated in the World Conference on Disaster Reduction in Kobe in 2005. Synergies between the frameworks for disaster risk management and measures to mainstream adaptation to climate change are increasingly being recognised (IFRC, 2002; Sperling and Szekely, 2005).

Box 5.2. **Attention to climate-development links in the IPCC Fourth Assessment Report**

The terms of reference for the IPCC Fourth Assessment Report give explicit and extensive consideration to the links between climate change and development. The implications of sustainable development are a cross-cutting issue to be covered in all chapters on regional and sectoral impacts of Working Group II. Development and mainstreaming issues also feature prominently in the synthesis chapters of Working Group II, which specifically address operational aspects of adaptation and issues related to mainstreaming of adaptation in development activities (Chapter 17), links among adaptation, mitigation and sustainable development (Chapter 18) and the broader regional and sectoral development implications of climate change and how they relate to access to resources and technology, as well as equity (Chapter 20).

The links between mitigation and development, meanwhile, will be considered in significant depth in the contribution of Working Group III. The influences of development trends and goals are considered an important issue related to mitigation in the long-term context, and will be examined in further detail in several chapters. Chapter 12 will also assess the literature on how sustainable development goals and mitigation policies inter-relate, and opportunities to link them further.

Source: Outlines of the contributions of Working Group II and III to the IPCC Fourth Assessment Report.

In addition to such *demand pull* from the development community, a much more visible *supply push* towards mainstreaming is now coming from the climate change community. The IPCC Third Assessment Report in 2001 brought the links between sustainable development and adaptation into much sharper focus within the climate change agenda. The Fourth Assessment Report, due in 2007, is seeking to deepen these links (Box 5.2) and will have a more explicit focus on operational aspects of adaptation.

A complex architecture of international funding sources for promoting adaptation in developing countries is also in place. It includes two funds under the UNFCCC – the Least Developed Countries Fund and Special Climate Change Fund – as well as the Adaptation Fund, which is under the Kyoto Protocol. The LDCF is already funding the development of NAPAs, while the SCCF is expected to promote sectoral initiatives on both adaptation and mitigation. Meanwhile, the entry into force of the Kyoto Protocol in February 2005 means the Clean Development Mechanism will move from an early implementation phase to full

BRIDGE OVER TROUBLED WATERS – ISBN 92-64-01275-3 – © OECD 2005

operation. The CDM will encourage investment in developing-country projects that limit emissions while promoting sustainable development. Part of its proceeds will be used to help put the Adaptation Fund into operation.

4. Five priorities for the road ahead

The above discussion shows that there has been considerable evolution in the dialogue between the climate change and development communities. What may have started with views akin to "climate-centrism" on the one hand and "why bother" on the other is slowly turning into a more nuanced exchange on whether and how to integrate climate change considerations into development efforts. Looking ahead, there are several opportunities to capitalise on this mutual reaching out to promote a broader agenda for effective integration of climate change considerations within development activities.

4.1. Making climate information more relevant and usable

It is imperative for decision makers in the development community to have access to credible and context-specific climate information as a basis for decisions on mainstreaming. That includes information on the cost and effectiveness of integrating adaptation or mitigation measures within development planning. Perhaps even more fundamental in the case of adaptation is information on the impacts of climate variability and change on particular development activities, including evaluation of associated uncertainty and information on the temporal and spatial scales over which the impacts might manifest themselves. While it might be naïve to call for a significant reduction in scientific uncertainty within climate model projections in the near future, it is important to recognise that development planners routinely face uncertainty when making decisions. Nevertheless, more can be done to facilitate transparent communication of the uncertainty (and the spatial/temporal scales) associated with climate variables of interest to particular decision makers. Concerning economic aspects, significant progress has been made on assessing the costs of mitigation and benefits of adaptation, but much is still needed on the benefits of mitigation and costs of adaptation (Corfee-Morlot and Agrawala, 2004). Analysis of the costs and distributional aspects of adaptation could also assist sectoral decision makers in determining the degree to which they should integrate such response measures within their core activities. This is an area of policy analysis where the OECD could play an important role.

4.2. Developing and applying screening tools

In addition to improving the quality of climate information, tools and approaches are needed to help screen development activities for climate risk and prioritise responses. Such tools and approaches include methodologies to assess the potential exposure of a broad range of development activities and investments to climate risks. They could also include more sophisticated screening tools at the project level to identify the key climatic variables of relevance to the project, how they might change under climate change and what the implications of such changes might be on the viability of the project. Such screening can be accomplished in a two step-process (Burton and van Aalst, 2004). When a project is proposed, an initial risk screening can be used to categorise a project as being at high, medium or low risk. For projects at substantial risk, a more comprehensive risk assessment could then be conducted, along with analysis of potential measures to reduce the risk and the associated cost. This information could serve as an input to the final project design and appraisal. The development of such screening tools is still at an early stage. Field-testing of such tools and their diffusion to a wide range of project settings could greatly advance the integration of climate risk information in development activities.

4.3. Identifying and using appropriate entry points for climate information

Identification of appropriate entry points for climate change information in planned development activities is greatly needed. At the strategic level, climate change can intersect with three areas of development co-operation: humanitarian aid, poverty reduction and economic development, and natural resource management (Eriksen and Næss, 2003). Vulnerability to multiple stresses and measures to reduce it form a key component of all three. Climate change may be one determinant of this vulnerability, which is shaped by a range of biophysical and socio-economic stresses. Adaptation to climate change can be integrated in broader programmes for vulnerability reduction. Climate change considerations can also be made part of programmes to promote sustainable livelihoods, capacity building and broader risk management in the context of poverty alleviation, natural resource management and humanitarian aid. Potential entry points for the use of climate information and for integrating adaptation include land use planning; design of early warning systems and disaster response strategies; and planning and infrastructure for coastal defences, urban drainage and water supply, hydroelectricity generation and flood control (Eriksen and Næss, 2003).

Environmental impact assessments (EIAs) could be another entry point for mainstreaming both mitigation and adaptation concerns. The implications of projects for GHG emissions could in principle be included in EIA checklists. EIAs consider the impact of a project or activity *on* the environment, however, not the impact of the environment on the viability of a project, as would be the case for climate change impacts. EIA guidelines might need to be broadened to include consideration of these impacts.

It is also important to embed climate change considerations within planning mechanisms, ensure that the responsibility for co-ordination and monitoring lies with a sufficiently influential department, foster cross-sectoral co-ordination and ensure that the process is country-driven. Furthermore, attention should be given not only to investment plans but also to policies and legislation. These priorities for mainstreaming of climate change are in fact similar to the guiding principles for good development planning as reflected, for instance, in the DAC Strategies for Sustainable Development guidelines (OECD, 2001).

4.4. *Shifting emphasis to implementation, as opposed to developing new plans*

A further need is to ensure that mainstreaming is not confined to simple "bean counting" of climate change action plans or mention of climate change in planning documents. In many instances, rather than requiring radically new responses, climate change might only reinforce the need for implementation of measures that already are, or should be, environmental or development priorities. Examples might include water or energy conservation, forest protection and afforestation, flood control, building of coastal embankments, dredging to improve river flow and protection of mangroves. Often such measures have already been called for in national and sectoral planning documents but not successfully implemented. Reiteration of the measures in elaborate climate change plans is unlikely to have much effect on the ground unless barriers to effective implementation of the existing sectoral and development plans are confronted.

In some cases the barriers might be institutional, such as sectoral segmentation within governments and development co-operation agencies. In other cases they might have to do with resource availability or the reliability of funding. In still other instances the real barriers to implementation might stem from a lack of policy coherence. Putting the spotlight on implementation, therefore, could put the focus on greater accountability in action on the ground.

4.5. *Meaningful co-ordination and sharing of good practices*

A key priority for the future is developing mechanisms to forge successful links between mainstreaming initiated under the UNFCCC and Kyoto Protocol and more bottom-up risk management initiatives by national and sectoral planners. A corollary link could be between activities initiated to achieve development objectives, such as the Millennium Development Goals, and more bottom-up consideration of the impacts of climate change.

Also needed is greater engagement of the private sector and local communities in mainstreaming efforts. Thus far the emphasis has been largely on efforts initiated by governments and international organisations.

Another priority that has not received sufficient attention is transboundary and regional co-ordination. Most climate change action and adaptation plans are at the national level while many impacts cut across national boundaries. Meaningful integration of a range of climate risks, from flood control to dry season flows to glacial lake hazards, would require greater co-ordination on data collection, monitoring and policies at the regional level. Responses to climate change could also have implications for more distant economies through mechanisms such as trade, which might also require co-ordination.

Finally, guidance on comprehensive climate risk management in development would facilitate policy coherence, allow for joint building of experience and promote sharing of tools and experiences across and between governments and development co-operation agencies. As the co-ordinating forum for international donors, the OECD could help the development co-operation and climate change communities find this kind of common ground. The DAC has already produced related guidance documents on EIA, disaster mitigation, strategies for sustainable development and the integration of the Rio conventions in development co-operation. The disaster mitigation document contains many elements relevant for climate risk management, and could be revisited with a specific focus on climate change adaptation. An important issue to be addressed when preparing such guidance is which elements of previous guidelines have been followed, which have been difficult to implement, and why. They could be identified, for example, through "backcast" studies of projects or plans that could have been more successful had the guidance been followed; by analysing best practice areas; and by examining the viability of climate risk management standards, for instance, as an extension of existing EIA safeguard requirements for development projects.

Perhaps the most fundamental message from this volume is that the climate change and development communities are not monolithic blocks that can be linked by a simple handshake. Rather, mainstreaming may require a meshing at multiple levels between the diverse range of actors and institutions connected with the two fields. Considerable progress has been made in this direction, but greater co-ordination is still needed at many levels, both within each domain and between them.

ISBN 92-64-01275-3
Bridge Over Troubled Waters
Linking Climate Change and Development

References

van Aalst, M. and S. Bettencourt (2004). "Vulnerability and Adaptation in Pacific Island Countries", in A. Mathur, I. Burton and M. van Aalst (eds.), *An Adaptation Mosaic: A Sample of Emerging World Bank Work in Climate Change Adaptation*, World Bank, Washington.

Abu-Zeid, M. and S. Abdel-Dayem (1992), "Egypt Programmes and Policy Options for Facing the Low Nile Flows", in M.A. Abu-Zeid and A.K. Biswas (eds.), *Climatic Fluctuations and Water Management*, Butterworths and Heinemann, Oxford, pp. 48-58.

Adger, W.N. *et al.* (2003), "Adaptation to Climate Change in the Developing World", *Progress in Development Studies*, Vol. 3, No. 3, pp. 179-95.

Agrawala, S. (2004), "Mainstreaming Adaptation in Development Planning and Assistance", presented at the OECD Global Forum on Sustainable Development: Development and Climate Change, Paris, 11-12 November.

Agrawala, S. and M. Berg (2002), "Development and Climate Change Project: Concept Paper on Scope and Criteria for Case Study Selection", COM/ENV/EPOC/DCD/DAC(2002)1/Final, OECD, Paris.

Agrawala, S. and M.A. Cane (2002), "Sustainability: Lessons from Climate Variability and Climate Change", *Columbia Journal of Environmental Law*, Vol. 27, No. 2, pp. 309-21.

Ahmed, A.U. (2002), "Reviewing the Policy Regime in Relation to Water Resources Vulnerability to Climate Change in Bangladesh", presented at the National Dialogue on Water and Climate, Dhaka, 12-14 December.

Ahmed, A.U. (2003), "Climate Change and Development in Bangladesh", consultant report for the OECD Development and Climate Change Project.

Ahmed, A.U. (2005), "Adaptation Options for Managing Water-related Extreme Events under Climate Change Regime: Bangladesh Perspectives", in M.M.Q. Mirza and Q.K. Ahmad (eds.), *Climate Change and Water Resources in South Asia*, Taylor and Francis, London/Leiden.

Ahmed, A.U. *et al.* (1998), "Vulnerability of Forest Ecosystems of Bangladesh to Climate Change", in S. Huq *et al.* (eds.), *Vulnerability and Adaptation to Climate Change for Bangladesh*, Kluwer Academic Publishers, Dordrecht, pp. 93-111.

Altmann, J. *et al.* (2002), "Dramatic change in local climate patterns in the Amboseli Basin, Kenya", African Journal of Ecology, Vol. 40, No. 3, pp. 248-51.

Baethgen, W.E. (1997), "Vulnerability of the agricultural sector of Latin America to climate change", *Climate Research*, Vol. 9, No. 1-2, pp. 1-7.

Baethgen, W.E. and D.L. Martino (2004), "Mainstreaming Climate Change Responses in Economic Development of Uruguay", presented at the OECD Global Forum on Sustainable Development: Development and Climate Change, Paris, 11-12 November, ENV/EPOC/GF/SD/RD(2004)2/FINAL, OECD, Paris.

Barthelet, P., L. Terray and S. Valcke (1998), "Transient CO_2 experiment using the ARPEGE/OPAICE non flux corrected coupled model", *Geophysical Research Letters*, Vol. 25, No. 13, pp. 2277-80.

Boville, B.A. and P.R. Gent (1998), "The NCAR Climate System Model, Version One", *Journal of Climate*, Vol. 11, pp. 1115-30.

Broad, K. and S. Agrawala (2000), "The Ethiopia Food Crisis: Uses and Limits of Climate Forecasts", *Science*, Vol. 289, pp. 1693-94.

Burton, I. and M. van Aalst (1999), "Come Hell or High Water: Integrating Climate Change Vulnerability and Adaptation into Bank Work", World Bank Environment Department Papers, No. 72, Climate Change Series, World Bank, Washington.

Burton, I. and M. van Aalst (2004), "Look Before You Leap: A Risk Management Approach for Incorporating Climate Change Adaptation into World Bank Operations", World Bank, Washington.

Church, J.A. *et al.* (2001), "Changes in Sea Level", in IPCC, *Climate Change 2001: The Scientific Basis*, Cambridge University Press, Cambridge, pp. 639-93.

Conway, D. (2002), "Extreme Rainfall Events and Lake Level Changes in East Africa: Recent Events and Historical Precedents", in E.O. Odada and D.O. Olago (eds.), *The East African Great Lakes: Limnology, Palaeolimnology and Biodiversity*, Advances in Global Change Research, Vol. 12, Kluwer Academic Publishers, Dordrecht, pp. 63-92.

Conway, D. (2005), "From headwater tributaries to international river basin: observing and adapting to climate variability and change in the Nile Basin", *Global Environmental Change*, Vol. 15, No. 2, pp. 99-114.

Conway, D. *et al.* (1996), "Future availability of water in Egypt: the interaction of global, regional and basin scale driving forces in the Nile Basin", *Ambio*, Vol. 25, No. 5, pp. 336-42.

Corfee-Morlot, J. and S. Agrawala (2004), "The Benefits of Climate Policy", *Global Environmental Change*, Vol. 14, No. 3, pp. 197-99.

Council of the European Union (2004), "Council Conclusions: Climate Change in the Context of Development Co-operation", *http://register.consilium.eu.int/pdf/en/04/st15/st15164.en04.pdf*.

Cubasch, U. *et al.* (2001), "Projections of future climate change", in IPCC, *Climate Change 2001: The Scientific Basis*, Cambridge University Press, Cambridge.

Department of Hydrology and Meteorology (2005), Tsho Rolpa GLOF Risk Reduction Project website, Ministry of Environment, Science and Technology, Government of Nepal, Kathmandu, *www.dhm.gov.np/tsorol/index.htm*.

Emori, S. *et al.* (1999), "Coupled ocean-atmosphere model experiments of future climate change with an explicit representation of sulfate aerosol scattering", *Journal of the Meteorological Society of Japan*, Vol. 77, No. 6, pp. 1299-1307.

Eriksen, S. and L.O. Næss (2003), *Pro-Poor Climate Adaptation – Norwegian Development Co-operation and Climate Change Adaptation: An Assessment of Issues, Strategies and Potential Entry Points*, CICERO Report 2003:02, Center for International Climate and Environmental Research, Oslo.

European Commission (2003), "Communication from the Commission to the Council and European Parliament: Climate Change in the Context of Development Co-operation", COM(2003)85FINAL, European Commission, Brussels, *http://europa.eu.int/eur-lex/en/com/cnc/2003/com2003_0085en01.pdf*.

Feresi, J. *et al.* (1999), "Climate Change Vulnerability and Adaptation Assessment for Fiji", draft, November.

Flato, G.M. *et al.* (2000), "The Canadian Centre for Climate Modelling and Analysis global coupled model and its climate", *Climate Dynamics*, Vol. 16, No. 6, pp. 451-67.

Gordon, C. *et al.* (2000), "The simulation of SST, sea ice extents and ocean heat transports in a version of the Hadley Centre coupled model without flux adjustments", *Climate Dynamics*, Vol. 16, No. 2-3, pp. 147-68.

Gordon, H.B. and S.P. O'Farrell (1997), "Transient climate change in the CSIRO coupled model with dynamic sea ice", *Monthly Weather Review*, Vol. 125, No. 5, pp. 875-907.

Halcrow and Associates (2001), "Options for the Ganges Dependent Area, Draft Final Report (Vol. 2)", report for Water Resources Planning Organization, Ministry of Water Resources, Dhaka.

Hay, S.I. *et al.* (2002), "Climate change and the resurgence of malaria in the East African highlands", *Nature*, Vol. 415, No. 6874, pp. 905-9.

Hemp, A. (2003), "Climate Impacts and Responses in Mount Kilimanjaro", consultant report for the OECD Development and Climate Change Project.

Hemp, A. (2005), "Climate change driven forest fires marginalize the impact of ice cap wasting on Kilimanjaro", *Global Change Biology*, Vol. 11, No. 7, pp. 1013-1023.

Hudson, R.A. and S.W. Meditz (eds.) (1990), *Uruguay, A Country Case Study*, Federal Research Division, Library of Congress, Washington.

Hulme, M. *et al.* (2000), *Using a Climate Scenario Generator for Vulnerability and Adaptation Assessments: MAGICC and SCENGEN Version 2.4 Workbook*, Climatic Research Unit, University of East Anglia, Norwich.

Huq, S. (2004), "International policy in supporting adaptation", *Insights* (quarterly journal of id21, Institute of Development Studies), No. 53, December, p. 3.

Huq, S. (2002), "Lessons Learned from Adapting to Climate Change in Bangladesh", submission to Climate Change Team, World Bank, *www.iied.org/docs/climate/lessons_oct02.pdf*.

Huq, S. *et al.* (eds.) (1999), *Vulnerability and Adaptation to Climate Change for Bangladesh*, Kluwer Academic Publishers, Dordrecht.

IEA (2002), *World Energy Outlook*, OECD/IEA, Paris.

IFRC (2002), *World Disasters Report 2002: Focus on Reducing Risk*, International Federation of Red Cross and Red Crescent Societies, Geneva.

IPCC (1995), *IPCC Second Assessment: Climate Change 1995*, Cambridge University Press, Cambridge.

IPCC (2001a), *Climate Change 2001: The Scientific Basis*, Cambridge University Press, Cambridge.

IPCC (2001b), *Climate Change 2001: Impacts, Adaptation and Vulnerability*, Cambridge University Press, Cambridge.

Ives, J.D. (1986), "Glacial Lake Outburst Floods and Risk Engineering in the Himalaya", ICIMOD Occasional Paper No. 5, International Centre for Integrated Mountain Development, Kathmandu.

JICA (1991), "Plan Quinquenal de Forestación Nacional de la República Oriental del Uruguay", Final Report, March, Japan International Co-operation Agency.

Johns, T.C. *et al.* (1997), "The second Hadley Centre coupled ocean-atmosphere GCM: Model description, spinup and validation", *Climate Dynamics*, Vol. 13, No. 2, pp. 103-34.

Jones, R. *et al.* (2004), "Assessing Future Climate Risks", in B. Lim and E. Spanger-Siegfried (eds.), *Adaptation Policy Framworks for Climate Change: Developing Strategies, Policies and Measures*, Cambridge University Press, Cambridge, pp. 119-43.

Kaser, G. *et al.* (2004), "Modern glacier retreat on Kilimanjaro as evidence of climate change: Observations and facts", *International Journal of Climatology*, Vol. 24, No. 3, pp. 329-39.

Klein, R.J.T. (2001), *Adaptation to Climate Change in German Official Development Assistance: An Inventory of Activities and Opportunities, with a Special Focus on Africa*, Deutsche Gesellschaft für Technische Zusammenarbeit (GTZ), Eschborn.

Koshy, K. and L. Philip (2002), "Capacity Enhancement for the Pacific", *Tiempo*, Vol. 45, No. 9, pp. 1-9.

Lal, P.N. (1990), "Conservation or Conversion of Mangroves in Fiji: An Ecological Economic Analysis", Occasional Paper 11, Environmental Policy Institute, East-West Center, Honolulu.

Leclainche, Y. *et al.* (2001), "The role of sea ice thermodynamics in the Northern Hemisphere climate as simulated by a global coupled ocean-atmosphere model", IPSL Note 21, October, Institut Pierre-Simon Laplace des sciences de l'environnement, Paris.

Liu, X. and B. Chen (2000), "Climatic Warming in the Tibetan Plateau During Recent Decades", *International Journal of Climatology*, Vol. 20, No. 14, pp. 1729-42.

Manabe, S. *et al.* (1991), "Transient responses of a coupled ocean-atmosphere model to gradual changes of atmospheric CO_2 – Part I: Annual mean response", *Journal of Climate*, Vol. 4, No. 8, pp. 785-818.

McAveney, B.J. *et al.* (2001), "Model Evaluation", in IPCC, *Climate Change 2001: The Scientific Basis*, Cambridge University Press, Cambridge, pp. 471-524.

Mool *et al.* (2002), *Inventory of Glaciers, Glacial Lakes and Glacial Lake Outburst Floods: Monitoring and Early Warning Systems the Hindu Kush-Himalayan Region: Nepal*, International Centre for Integrated Mountain Development, Kathmandu.

Munasinghe, M. (2002), *Analysing the nexus of sustainable development and climate change: An overview*, COM/ENV/EPOC/DCD/DAC(2002)2/FINAL, OECD, Paris.

Nakicenovic, N. and R. Swart (eds.) (2000), *Emissions Scenarios: A Special Report of Working Group III of the Intergovernmental Panel on Climate Change*, Cambridge University Press, Cambridge.

Nunn, P. *et al.* (1993), *Assessment of Coastal Vulnerability and Resilience to Sea Level Rise and Climate Change, Case Study – Viti Levu Island, Fiji, Phase 1: Concepts and Approach*, Technical Report, South Pacific Regional Environment Programme, Apia.

OECD (2000), DAC Statistical Reporting Directives, *www.oecd.org/dac/stats/dac/directives*.

OECD (2001), *The DAC Guidelines – Strategies for Sustainable Development: Guidance for Development Co-operation*, OECD, Paris.

OECD (2003a), *Development and Climate Change in Nepal: Focus on Water Resources and Hydropower*, COM/ENV/EPOC/DCD/DAC(2003)1/FINAL, OECD, Paris.

BRIDGE OVER TROUBLED WATERS – ISBN 92-64-01275-3 – © OECD 2005

OECD (2003b), *Development and Climate Change in Bangladesh: Focus on Coastal Flooding and the Sundarbans*, COM/ENV/EPOC/DCD/DAC(2003)3/FINAL, OECD, Paris.

OECD (2003c), *Development and Climate Change in Fiji: Focus on Coastal Mangroves*, COM/ENV/EPOC/DCD/DAC(2003)4/FINAL, OECD, Paris.

OECD (2003d), *Development and Climate Change: Focus on Mount Kilimanjaro*, COM/ENV/EPOC/DCD/DAC(2003)5/FINAL, OECD, Paris.

OECD (2003e), List of Aid Recipients (as of 1 January 2003), *http://www.oecd.org/dataoecd/35/9/2488552.pdf.*

OECD (2004a), *Development and Climate Change in Egypt: Focus on Coastal Resources and the Nile*, COM/ENV/EPOC/DCD/DAC(2004)1/FINAL, OECD, Paris.

OECD (2004b), *Development and Climate Change in Uruguay: Focus on Coastal Zones, Agriculture and Forestry*, COM/ENV/EPOC/DCD/DAC(2004)2/FINAL, OECD, Paris.

OECD (2004c), International Development Statistics Online Databases, *www.oecd.org/dac/stats/idsonline.*

Pócs, T. (1976), "The Role of the Epiphytic Vegetation in the Water Balance and Humus Production of the Rain Forests of the Uluguru Mountains, East Africa", *Boissiera*, Vol. 24, pp. 499-503.

Power, S.B. *et al.* (1998), *A Coupled General Circulation Model for Seasonal Prediction and Climate Change Research*, BMRC Research Report No. 66, Bureau of Meteorology Research Centre, Melbourne.

Rahman, A. and M. Alam (2003), *Mainstreaming Adaptation to Climate Change in Least Developed Countries (LDCs) – Working Paper 2: Bangladesh Country Case Study*, International Institute for Environment and Development, London.

Raksakulthai, V. (2003), "Nepal's Hydropower Sector: Climate Change, GLOFs, and Adaptation", consultant report for the OECD Development and Climate Change Project.

Rana, B. *et al.* (2000), "Hazard Assessment of the Tsho Rolpa Glacier Lake and Ongoing Remediation Measures", *Journal of Nepal Geological Society*, Vol. 22, pp. 563-70.

Raper *et al.* (1996), "Global Sea-Level Rise: Past and Future", in J.D. Milliman and B.U. Haq (eds.), *Sea-level Rise and Coastal Subsidence*, Kluwer Academic Publishers, Dordrecht, pp. 11-45.

Risbey, J.S. *et al.* (2002), "Exploring the Structure of Regional Climate Scenarios by Combining Synoptic and Dynamic Guidance and GCM Output", *Journal of Climate*, Vol. 15, No. 9, pp. 1036-50.

Roeckner, E. *et al.* (1996), "The Atmospheric General Circulation Model ECHAM4: Model Description and Simulation of Present-Day Climate", MPI Report No. 218, Max-Planck-Institut für Meteorologie, Hamburg.

Russell, G.L., J.R. Miller and D. Rind (1995), "A coupled atmosphere-ocean model for transient climate change studies", *Atmosphere-Ocean*, Vol. 33, No. 4, pp. 683-730.

Santer, B.D. *et al.* (1990), "Developing Climate Scenarios from Equilibrium GCM Results", MPI Report No. 47, Max-Planck-Institut für Meteorologie, Hamburg.

Sarmett, J.D. and S.A. Faraji (1991), "The Hydrology of Mount Kilimanjaro: An Examination of Dry Season Runoff and Possible Factors Leading to its Decrease", in W.D. Newmark (ed.), *The Conservation of Mount Kilimanjaro*, IUCN, Gland, pp. 53-70.

Shackley, S. and B. Wynne (1995), "Integrating Knowledges for Climate Change: Pyramids, Nets and Uncertainties", *Global Environmental Change*, Vol. 5, No. 2, pp. 113-26.

Shakya, N.M. (2003), "Hydrological Changes Assessment and Its Impact on Hydro Power Projects of Nepal", in draft proceedings of the Consultative Workshop on Climate Change Impacts and Adaptation Options in Nepal's Hydropower Sector with a Focus on Hydrological Regime Changes Including GLOF, Department of Hydrology and Meteorology and Asian Disaster Preparedness Center, 5-6 March, Kathmandu.

Shrestha, A.B. *et al.* (1999), "Maximum Temperature Trends in the Himalaya and Its Vicinity: An Analysis Based on Temperature Records from Nepal for the Period 1971-94", *Journal of Climate*, Vol. 12, No. 9, pp. 2775-89.

Shrestha, M.L. and A.B. Shrestha (2004), "Recent Trends and Potential Climate Change Impacts on Glacier Retreat/Glacier Lakes in Nepal and Potential Adaptation Measures", presented at the OECD Global Forum on Sustainable Development: Development and Climate Change, Paris, 11-12 November, ENV/EPOC/GF/SD/RD(2004)6/FINAL, OECD, Paris.

Shukla, P.R., M. Kapshe and A. Garg (2004), "Development and Climate: Impacts and Adaptation for Infrastructure Assests in India", presented at the OECD Global Forum on Sustainable Development: Development and Climate Change, Paris, 11-12 November.

Smit, B. *et al.* (2001), "Adaptation to Climate Change in the Context of Sustainable Development and Equity", in IPCC, *Climate Change 2001: Impacts, Adaptation, and Vulnerability*, Cambridge University Press, Cambridge, pp. 877-912.

Smith, J.B. *et al.* (1998), "Considering Adaptation to Climate Change in the Sustainable Development of Bangladesh", report to the World Bank by Stratus Consulting Inc., Boulder.

Smith, J.B. *et al.* (2003), "MAGICC/SCENGEN Analysis of Climate Change Scenarios for Bangladesh, Egypt, Fiji, Nepal, Tanzania and Uruguay", consultant report for the OECD Development and Climate Change Project, Stratus Consulting Inc., Boulder.

Sperling, F. (ed.) (2003), *Poverty and Climate Change: Reducing the Vulnerability of the Poor through Adaptation*, report by the African Development Bank, Asian Development Bank, UK Department for International Development (UK), Federal Ministry for Economic Co-operation and Development (Germany), Ministry of Foreign Affairs – Development Co-operation (Netherlands), OECD, United Nations Development Programme, United Nations Environment Programme and World Bank.

Sperling, F. and F. Szekely (2005), "Disaster Risk Management in a Changing Climate", informal discussion paper prepared for the World Conference on Disaster Reduction on behalf of the Vulnerability and Adaptation Resource Group, Washington.

Strzepek, K.M. *et al.* (1995), "An Assessment of Integrated Climate Change Impacts on Egypt", in K.M. Strzepek and J.B. Smith (eds.), *As Climate Changes: International Impacts and Implications*, Cambridge University Press, Cambridge.

Swart, R. *et al.* (2003), "Climate Change and Sustainable Development: Expanding the Options", *Climate Policy*, Vol. 3, Supplement 1, pp. S19-40.

Tokioka, T. *et al.* (1996), "A Transient CO_2 Experiment with the MRI CGCM: Annual Mean Response", CGER's Supercomputer Monograph Report Vol. 2, Center for Global Environmental Research, National Institute for Environmental Studies, Environment Agency of Japan, Ibaraki.

UNFCCC (2005), List of national focal points (as of January 2005), *http://maindb.unfccc.int/public/nfp.pl*, United Nations Framework Convention on Climate Change, Bonn.

Uruguay (2002), "Estudio de Apoyo a la Aplicación del Mecanismo para el Desarrollo Limpio del Protocolo de Kioto en Uruguay", Ministerio de Vivienda, Ordenamiento Territorial y Medio Ambiente, Montevideo.

Voss, R., R. Sausen and U. Cubasch (1998), "Periodically synchronously coupled integrations with the atmosphere-ocean general circulation model ECHAM3/LSG", *Climate Dynamics*, Vol. 14, No. 4, pp. 249-66.

Washington, W.M. and G.A. Meehl (1996), "High-latitude climate change in a global coupled ocean-atmosphere-sea ice model with increased atmospheric CO_2", *Journal of Geophysical Research*, Vol. 101, No. D8, pp. 12795-801.

Washington, W.M. *et al.* (2000), "Parallel Climate Model (PCM) control and transient simulations", *Climate Dynamics*, Vol. 16, No. 1°-11, pp. 755-74.

Waterbury, J. (2002), *The Nile Basin: National Determinants of Collective Action*, Yale University Press, New Haven.

Wichelns, D. (2002), "Economic Analysis of Water Allocation Policies Regarding Nile River Water in Egypt", *Agricultural Water Management*, Vol. 52, No. 2, pp. 155-75.

World Bank (2000a), *Bangladesh: Climate Change and Sustainable Development*, Report No. 21104 BD, World Bank South Asia Rural Development Unit, Dhaka.

World Bank (2000b), *Cities, Seas, and Storms: Managing Change in Pacific Island Economies – Volume IV: Adapting to Climate Change*, World Bank, Washington.

World Bank (2002), *World Development Indicators*, CD-ROM, World Bank, Washington.

Zhang, X.H. *et al.* (eds.) (2000), *IAP Global Atmosphere-Land System Model*, Science Press, Beijing.

OECD PUBLICATIONS, 2, rue André-Pascal, 75775 PARIS CEDEX 16
PRINTED IN FRANCE
(97 2005 09 1 P) ISBN 92-64-01275-3 – No. 54271 2005